Science in the Medieval World

History of Science Series, No. 5

SCIENCE IN THE MEDIEVAL WORLD

"Book of the Categories of Nations"

by
Ṣāʿid al-Andalusī

Translated and Edited by
Semaʿan I. Salem and Alok Kumar

University of Texas Press, Austin

Printed in the United States of America

First paperback printing, 1996

The paper used in this publication meets the minimum requirements of American National Standard for Information Sciences—Permanence of Paper for Printed Library Materials, ANSI Z39.48-1984.

Library of Congress Cataloging-in-Publication Data

Andalusī, Ṣāʿid ibn Aḥmad, 1029–1070.
[Ṭabaqāt al-ʾumam. English]
Science in the medieval world : book of the Categories of nations by Ṣāʿid al-Andalusī ; translated and edited by Semaʿan. I. Salem and Alok Kumar.—1st ed.
p. cm.—(History of science series ; no. 5)
Translation of: Ṭabaqāt al-ʾumam.
Includes bibliographical references and index.
ISBN 0-292-71139-5 (cloth)
ISBN 0-292-70469-0 (pbk.)
1. Science, Medieval. 2. Science—Islamic countries—History. I. Salem, Semaʿan I., 1927– . II. Kumar, Alok, 1954– . III. Title. IV. Series: History of Science series (Austin, Tex.) ; no. 5.
Q124.97.A5313 1991
509'.02—dc20
91-13332
CIP

The cover illustration is taken from a map of the "known world" (the Eastern Hemisphere) by al-Idrisi, the first geographer to portray the world on a disk, according to al-Andalusī.

Contents

ACKNOWLEDGMENTS VII

SYSTEM OF TRANSLITERATION IX

ṢĀʿID AL-ANDALUSĪ XI

INTRODUCTION XV

Ṭabaqāt al-ʾUmam

CHAPTER 1. The Seven Original Nations 3

CHAPTER 2. The Two Categories of Nations 6

CHAPTER 3. Nations Having No Interest in Science 7

CHAPTER 4. Nations That Cultivated the Sciences 9

CHAPTER 5. Science in India 11

CHAPTER 6. Science in Persia 15

CHAPTER 7. Science of the Chaldeans 18

CHAPTER 8. Science in Greece 20

CHAPTER 9. Science of the Romans 31

CHAPTER 10. Science in Egypt 35

CHAPTER 11. The Arabs: General Information 38

CHAPTER 12. Science in the Arab Orient 46

CHAPTER 13. Science in al-Andalus 58

CHAPTER 14. Science of Banū Israel 79

NOTES 83

BIBLIOGRAPHY 105

INDEX 109

Acknowledgments

We are very grateful to J. G. Salem and Professor L. S. Lerner for editing parts of the manuscript and commenting on its contents. Our sincere thanks to Sid Sims, Cathrine L. Ida, and Sharlene La Forge for their assistance in the library search and the procurement of documents, and to Professor B. L. Scott for his help with computer programming. Although they have not participated in this project, Alok Kumar is grateful to G. S. Sarswat and Shanti Devi for their constant support and encouragement.

System of Transliteration

(adapted from *The Concise Encyclopedia of Islam*, 1989)

Transliteration	*Arabic Letter*	*Transliteration*	*Arabic Letter*
ṭ	ط	ʾ	ء
ẓ	ظ	b	ب
ʿ	ع	t	ت
gh	غ	th	ث
f	ف	j	ج
q	ق	ḥ	ح
k	ك	kh	خ
l	ل	d	د
m	م	dh	ذ
n	ن	r	ر
ah/at	ه	z	ز
w	و	s	س
y	ي	sh	ش
t	ة/ـة	ṣ	ص
		ḍ	ض

Short Vowels

a	َ	(e.g., ba)	بَ
u	ُ	(e.g., bu)	بُ
i	ِ	(e.g., bi)	بِ

Long Vowels

ā	اَ	(e.g., bā)	بَا
ū	وُ	(e.g., bū)	بُو
ī	يِ	(e.g., bī)	بِي

Ṣāᶜid al-Andalusī

Abū al-Qāsim Ṣāᶜid ibn Abū al-Walīd Aḥmad ibn ᶜAbd al-Raḥmān ibn Muḥammad ibn Ṣāᶜid ibn ᶜUthmān al-Taghlibi al-Qūrtūbi, better known as Ṣāᶜid al-Andalusī[1] or Qāḍi [Judge] Ṣāᶜid, was born in Almuriyyah [Almería] in southern Spain in A.D. 1029/A.H. 420.[2] He was a philologist, natural philosopher, and historian as well as judge. As his name indicates, he was a member of the tribe of Taghlib, one of the strongest and largest tribes of Arabia.[3] Some of its members entered Spain during the Arab invasion of that country in A.D. 711/A.H. 92 and prospered there.[4]

Ṣāᶜid was born to a well-to-do family, whose members spent their time and wealth in quest of knowledge and education. His grandfather, nicknamed Abū al-Muṭarraf ᶜAbd al-Raḥmān, was a judge in Sidonia, Spain. He secured this position after returning from a voyage to the Arab Orient,[5] presumably in recognition of the education he had acquired there. He later relinquished this prestigious position to spend all his time in the pursuit of science and learning. Ṣāᶜid's father also occupied a highly respected position in the city of Córdoba, hence the name Qūrtūbi. This is where Ṣāᶜid received his early education. As an adolescent, he toured Muslim Spain to further his education. At the age of seventeen, he moved to the city of Toledo, then known by its Arabic name, Ṭulayṭilah (Blachère 1935 : 8).

Our knowledge of the life of the young scholar prior to this time is somewhat vague, and the reason behind his move to Toledo is not known. It is possible that he intended to further his education: Toledo was then a well-established literary center for science and education, and Córdoba had not then completely recovered from the wanton destruction it had suffered early in the eleventh century. Moreover, he could have gone to study with a particular scholar (Bū-ᶜAlwan 1985 : 13).

Toledo was then the capital of the princedom of Banū al-Nūn. This princedom was established by Ismāᶜīl al-Ẓāffir in A.D. 1035/A.H. 426.[6] When Ṣāᶜid entered the city, it was governed by Yaḥyā ibn Dhi al-Nūn,[7] son of al-Ẓāffir. This prince reigned from A.D. 1037 to 1074/A.H. 428 to 466 and extended his kingdom[8] to include Valencia and most of the eastern parts of Andalusia. His princedom became a political as well as an intellectual link between the northern and southern parts of Spain. Yaḥyā was succeeded by his grandson, al-

Qādir ibn Dhi al-Nūn, who lacked his grandfather's ability to govern. Al-Qādir fled from Toledo in A.D. 1080/A.H. 472 and the city was captured by Alfonso VI in May, A.D. 1085/A.H. 477.

As a result of Yaḥyā's efforts, Toledo became an important literary and intellectual center in which the various areas of knowledge were well represented. In addition to several authorities in Islamic science (*ḥadīth,* law, theology, tradition, etc.), there were mathematicians, such as Abū al-Walīd Hishām ibn Hishām ibn Khālid al-Kinānī, better known as Abū Walīd or al-Waqshi, and Abū Isḥaq ibn Ibrahim ibn Lub ibn Idris al-Tajibī; geometers, such as Muḥammad ibn Khyrah al-ᶜAṭṭar and Abū Jaᶜfar Aḥmad ibn Khamīs ibn ᶜĀmir ibn Damj, who was also a well-known astronomer; the astrologer Abū Bakr ibn Yaḥyā ibn Aḥmad, known as ibn al-Khayyaṭ [the tailor]; and several men of medicine, among them Abū ᶜUthmān Saᶜīd ibn Muḥammad ibn al-Bajunis and the vizier Abū al-Muṭarraf ᶜAbd al-Raḥmān ibn Muḥammad ibn ᶜAbd al-Kabīr ibn Yaḥyā ibn Wāfid ibn Muḥammad al-Lakhmī. All of these scholars are mentioned by Ṣāᶜid in *Ṭabaqāt al-ᵓUmam.*

One of the status symbols of kings and rulers of the time was the number and the quality of the scholars, philosophers, and poets in their courts. Probably for that reason, Yaḥyā ibn Dhi al-Nūn invited Ṣāᶜid into his service and appointed him a judge (hence the name Qāḍi Ṣāᶜid, or Judge Ṣāᶜid).

For their education, students flocked around the established scholars and thinkers of their time. Ṣāᶜid was no different. Like most Arab students, he chose to study law, Islamic religion, Arabic language, and Arabic literature. But later in life he specialized in the exact sciences, excelling in mathematics and astronomy. Among his most famous teachers was Abū Muḥammad al-Qāsim Abū al-Fatiḥ Muḥammad ibn Yūsuf,[9] specialist in Islamic religion, the Arabic language, and law.

In Toledo, Ṣāᶜid studied with Hishām al-Kinānī, known as Abū Walīd or al-Waqshi (referring to the name of the city of his birth, Huecas, a suburb of Toledo).[10] Abū Walīd was a well-established author and poet. He was also a *qāḍi* and a man of literature. He exercised considerable influence in the court of Yaḥyā ibn Dhi al-Nūn and introduced Ṣāᶜid to Yaḥyā's court.

Ṣāᶜid also studied with another well-established scholar, Abū Isḥaq Ibrahim ibn Idris al-Tajibī,[11] who taught him mathematics and observational astronomy.

While in the service of Yaḥyā, Ṣāᶜid divided his time among his duties as judge, his teaching, and his research in the history of science, theology, and literature. It was during this period that he wrote most of his work.

Ṣāᶜid al-Andalusī wrote several manuscripts, of various lengths, on a variety of subjects. All of his works except for *Ṭabaqāt al-ʾUmam* are lost. Many of the works of the Muslims in Spain met a similar fate. What we know about their achievements comes from the documents that were translated into Latin or were taken out of Spain prior to the Christian conquest.

Ṣāʿid wrote *Ṭabaqāt al-ʾUmam* in A.D. 1068/A.H. 460, when he was almost forty years old. By then, teaching was taking most of his time. It is believed that he discussed the content of this manuscript with some of his students; outstanding among them was Abū Bakr ʿAbd al-Bāqi ibn Muḥammad ibn Baryal, who was in part responsible for its style and language. Ṣāʿid was well read and had traveled extensively in Spain, which was then a leader in the scientific world. In writing *Ṭabaqāt al-ʾUmam,* he relied heavily on his personal contacts with other scholars and travelers; this can be deduced from the manuscript itself. Ṣāʿid writes with serene impartiality even about the scholars with whom he had personal contact. He shows great admiration for Aristotle and Ptolemy and is somewhat critical of the work of Abū Yūsuf Yaʿqūb ibn Isḥaq al-Kindi, the philosopher of the Arabs, and of the work of Abū Jaʿfar Muḥammad ibn Mūsā al-Khuwarizmi, the mathematician of the Arabs.

In addition to *Ṭabaqāt al-ʾUmam,* Ṣāʿid wrote a document on the art of astronomical observation, one on the history of nations, and another on the religions of nations. All are mentioned in *Ṭabaqāt al-ʾUmam.* He may have written a manuscript on the history of al-Andalus and one on the history of Islam.

Ṣāʿid al-Andalusī died in July, A.D. 1070/Shawwal, A.H. 462. Yaḥyā ibn Ṣāʿīd ibn Aḥmad ibn Yaḥyā al-Ḥadidī, the most illustrious dignitary in the court of ibn Dhi al-Nūn, read his official obituary.

Introduction

Although *Ṭabaqāt al-ʾUmam* is by far the most common Arabic title given to this work, other titles or variations on this title have frequently been used. Among them are *al-Taʿrif bi-Ṭabaqāt al-ʾUmam* [A Knowledge of . . .] and *Tārikh al-ʾUmam* [History of Nations]. Yaqūt (d. A.D. 1229/A.H. 626) and ibn Khallikan (d. A.D. 1282/A.H. 681) extracted passages from Ṣāʿid's book and referred to it as *Kitāb Akhbār al-Ḥukamāʾ* [Book of the Annals of Sages]. Ibn Saʿīd (d. A.D. 1286/A.H. 685) extracted exact passages from *Ṭabaqāt al-ʾUmam* and referred to it as *al-Taʿrīf bi-Akhbār Ḥukamāʾ al-ʾUmam Min al-ʿArab wa al-ʿAjam* [A Knowledge of the Annals of the Sages of the Nations from the Arabs and Non-Arabs]. It is quite possible that this was the original title of Ṣāʿid's book, later abbreviated to *Ṭabaqāt al-ʾUmam.* Varying titles for books were not uncommon in the Middle Ages and the Renaissance, and Ṣāʿid's work, becoming well known, could have been reproduced under different titles.

The author's name, صاعد, sounds more like Ṣawʿid than Saʿīd, which is a different Arabic proper name. But his name was written as Ṣāʿid in the French translation, the *Dictionary of Scientific Biography,* and other reference books. We have chosen tradition over phonetics.

The first modern study of this manuscript appeared in 1912. In 1907, the Jesuit Father Louis Shaykhū [Cheikho], bought a copy of the manuscript in Damascus and brought it to Beirut. He described it as being no more than two hundred years old and containing many errors. Later he was able to obtain a photocopy of the complete manuscript, as well as copies of segments of the manuscript, from the London Library. Based on these copies, he introduced some corrections, added appendices and footnotes, and published it under the title *Kitāb Ṭabaqāt al-ʾUmam* in 1912.

Ṭabaqāt al-ʾUmam[1] was translated into French by Régis Blachère in 1935 and published under the title *Kitâb Tabakât al-Umam (Livre des catégories des nations).*[2] It has become known in the West by that title. To avoid confusion, we have chosen to retain the familiar title *Categories of Nations,* though a more appropriate translation of the original title would probably be *Classifications of Nations.*

But in conformity with accepted English-Arabic transliteration, we have chosen to spell *ṭabaqāt* with a *q* rather than a *k*. The word *ṭabaqāt* is used frequently in the text; to preserve its full meaning, we retain it as such in the translation.[3]

In 1967, M. Baḥr al-ᶜUlūm edited and published *Ṭabaqāt al-ʾUmam* in Arabic. Recently, Ḥayat Bū-ᶜAlwan carried out a more comprehensive study of several copies of the manuscript, including the work of Shaykhū [Cheikho] and Blachère, and republished *Ṭabaqāt al-ʾUmam* in Arabic in 1985.

It is quite easy to make mistakes in copying Arabic script; for example, بـ تـ ثـ نـ يـ are five different letters of the Arabic alphabet. It is not uncommon to have two Arabic words that differ by only one dot. Thus one single dot could change the meaning of an entire sentence or make it incomprehensible. For our translation, we have thoroughly reviewed three of the four published versions. We have eliminated many mistakes and pointed out only serious errors and statements whose authenticity we could not verify.

As is always true of medieval manuscripts, the half-dozen or so existing copies of Ṣāᶜid's work have been reproduced at various times by different copyists.[4] Thus they differ in many respects. Dates, names, and places are not the same in all copies. As was common practice, some sections may have been added to the original, while others may have been intentionally or unintentionally deleted. For example, the copy at the Chester Beatty Library in Dublin, Ireland, ends with the Arab nation; the contribution of the nation of Israel is completely omitted from this copy.[5] Most of the copies are not dated; therefore, it is difficult to trace them back in time and to determine their accuracy relative to the original copy.

While many of the eleventh-century manuscripts were lost or destroyed, *Ṭabaqāt al-ʾUmam* survived primarily because of its relative importance and because of the interest it generated in scholarly circles. A small book that summarizes scientific contributions from all the nations of the world was surely considered a prize possession.

Ṣāᶜid spent most of his last few years conducting observations and teaching; a man of means and consequence, he organized a school of thought and gathered around him a group of students. Through his generous support and patronage, he freed them from the need to earn a living and enabled them to devote all their time to scholarly pursuits.[6] Ṣāᶜid's school survived him and continued to study his writings for several generations. One of his students, Abū Bakr ᶜAbd al-Bāqi ibn Muḥammad, also known as ibn Baryal al-Anṣāry, explicated *Ṭabaqāt al-ʾUmam* and improved on its style and language before handing it down to two of his own students, Abū Muḥammad ᶜAbd al-Haq ibn ᶜAṭiyah (d. A.D. 1147/A.H. 541) and Qāḍi ibn Abū ᶜĀmir ibn Shruyah (d. A.D. 1153/A.H. 548). These and several other scholars of that school referred to *Ṭabaqāt al-ʾUmam* in many of their writings. By so doing they established the book as a reference source for scientific information and to a certain extent ensured its survival.

Al-Zarqāli [Arzachel] Abū Isḥaq Ibrahim ibn Yaḥyā al-Naqqāsh was probably the most accomplished of Ṣāʿid's students.[7] He combined theoretical knowledge and technical skill to construct most of the astronomical instruments used in their observations. Al-Zarqāli also constructed the famed water clocks of Toledo, which in reality constituted a very precise lunar calendar. With guidance and assistance from his mentor, he also revised the astronomical tables of al-Khuwarizmi and prepared what became known as the Toledan Tables.[8] These highly successful tables became the basis for the Marseilles Tables, which were used throughout Europe by the twelfth century.[9]

Ṭabaqāt al-ʾUmam deals with practically all aspects of human knowledge; it discusses philosophy, religion, political science, geometry, mathematics, natural sciences, the arts, poetry, geography, climate, languages, and human behavior.[10] These various fields were referred to collectively as *ʿulūm* [sciences]. The book is organized in sections, each of which gives the contributions of one of the various nations. In certain cases, it discusses these contributions in some detail, stating the names of the scholars, the places of their birth, and with whom they studied. It contains the names of a large number of scientists and scholars and the dates of their contributions to the various branches of human intellect. The work is unique in the sense that it is comprehensive and highly condensed,[11] containing a tremendous amount of information. This made it a very desirable reference book requiring a relatively short time to copy. Thus several copies must have been made, and a few of them have survived.

In the first two centuries after its publication, it was quoted and referred to by several well-known scholars. Among them are Abū al-Qāsim Khalaf ibn ʿAbd al-Malik ibn Bashkuwal (d. A.D. 1183/A.H. 579), Abū ʿAbd Allah Muḥammad ibn ʿAbd Allah ibn al-Abbār (d. A.D. 1260/A.H. 658), Abū al-Ḥasan ʿAli ibn Yūsuf al-Qifṭī (d. A.D. 1246/A.H. 644), Muwaffaq al-Dīn Abū al-ʿAbbās Aḥmad ibn al-Qāsim ibn Abū ʿUṣaybiʿah (d. A.D. 1270/A.H. 668), and Abū al-Faraj Yūḥanna ibn al-ʿIbrī (d. A.D. 1268/A.H. 666). The world traveler Muḥammad ibn ʿAbd Allah ibn Baṭṭuṭah, writing about the origin of science in Egypt, copied verbatim from *Ṭabaqāt al-ʾUmam.*[12]

Interest in *Ṭabaqāt al-ʾUmam* did not subside for many centuries after its publication. Long after its science had been superseded, it was prized for its historic value. In A.D. 1632/A.H. 1042 al-Maqqarī made full use of the information in *Ṭabaqāt al-ʾUmam* while discussing eleventh-century scientific progress in Andalusia. In A.D.

1657/ A.H. 1067 Ḥaji Khalifah relied heavily on Ṣāʿid's book in writing about advances in astronomy during that period.

Early in the twelfth century, ʿAbd Allah ibn Muḥammad ibn Marzūq al-Yaḥsubi, a student of ibn Baryal, carried a copy of the *Ṭabaqāt* with him while on *ḥajj* to Makkah [Mecca]. On his way back he passed through Alexandria, Egypt, where he shared the book with the traditionalist Abū Ṭāhir al-Ṣalafī al-Miṣri [the Egyptian] (d. A.D. 1181/A.H. 576), who publicized it not only in Egypt but in the entire Middle East. Probably this, more than any other event, contributed to the preservation of *Ṭabaqāt al-ʾUmam;* most of the works of the Arabs in Spain perished or were destroyed during the reign of al-Ḥājib Abū ʿĀmir and after the Arabs were forced out of that country. "Ṭabaqāt al-Umam is important in two main respects: First, it gives an insight into the origin and cultivation of the sciences as they were known by the Andalusians of the eleventh century, and second, it enables us to gauge the extent of their cultivation and appreciation on Andalusian soil."[13] It is not only considered a fundamental work, but is also regarded as a very precise source of information. Probably for that reason George Sarton, the noted historian, writes, "Critical edition of the Ṭabaqāt al-ʾUmam. . . . An English version is very desirable."[14]

In his recent study, L. Richter-Bernburg mentions the shortcomings of Blachère's edition and the Iraqi edition and suggests that "a truly critical edition [of *Ṭabaqāt al-ʾUmam*] has long since been overdue."[15]

Franz Rosenthal, in his translation of ibn Khaldūn's *al-Muqaddimah,* used *Ṭabaqāt al-ʾUmam* as an authoritative source.[16] B. B. Lawrence, writing about Indian religions, states, "Qāḍi Ṣāʿid al-Andalusī (d. 1070 A.D.) notes, in Ṭabaqāt al-ʾUmam, that the Hindus have two sects; some are followers of Brahma, the others are Sabians."[17] M. S. Khan relies heavily on *Ṭabaqāt al-ʾUmam* in writing about ancient Indian sciences and culture.[18] M. Mahdi describes *Ṭabaqāt al-ʾUmam:* "In Muslim Spain . . . Ṣāʿid (d. 1070 A.D.) wrote his Classes of Nations . . . in which he presented the history, learning, character and social life of various nations. . . . He considered the cultivation of the sciences . . . a decisive moment in human history, and divided the nations, accordingly, into two broad categories."[19] And R. K. Chaube uses long quotations from Ṣāʿid's work to describe India as viewed by the Muslims.[20]

Ṭabaqāt al-ʾUmam is still being used as a precise reference to identify natural philosophers of the Islamic era; A. I. Sabra, writing in the *Dictionary of Scientific Biography* (*DBS*), uses this document to discuss the work of ʿAbd Allah ibn Muḥammad ibn Jaʿfar al-Farghānī (*DSB,* IV, 542), and to ascertain the presence of al-Ḥasan ibn al-Ḥasan ibn al-Haytham in Egypt in A.D. 1039/A.H. 430 (*DSB,* VI,

189). Similarly, David Pingree makes use of *Ṭabaqāt al-ʾUmam* to describe the life of ʿUmar ibn Farkhān al-Ṭabarī in the Abbāsid court (*DSB*, XIII, 538) and to explain some of the stories attributed to the Indian philosopher Kanka (*DSB*, VII, 223).

In this book, Ṣāʿid divides the people of the world into two classes. In one class he includes the eight nations that made significant contributions to science: the Indians, the Persians, the Chaldeans, the Greeks, the Romans, the Egyptians, the Arabs, and the Hebrews. In the second class he groups all those who did not cultivate science: the Chinese, the Turks, all the people of Africa (except Egypt), the people of northern Siberia, and the people of parts of the Balkan region. One is inclined to believe that geographical distances and language barriers played a significant role in this grouping and in Ṣāʿid's knowledge of the contributions of the various nations. He was very candid about this shortcoming; on many occasions he writes, "I have received no information about . . ." or "I have not been informed."

Of the nations that did not cultivate science, the Chinese and the Turks were the most advanced. These two nations excelled in manual labor, industrial technology, and the construction of various implements. But Ṣāʿid believes that the elites among God's creatures are those who have cultivated science, and he provides many examples where animals surpass humans in the performance of most other functions.

In discussing the contributions of the various nations, Ṣāʿid begins each section by describing the general traits of the people about whom he is writing. He briefly mentions their religion, their language, and their history. The geographical location of each nation is then presented in some detail. This is generally followed by the enumeration of the nation's celebrated scholars and a few statements about their lives and their works.

Ṣāʿid states that India is the first nation to have cultivated science. He praises the knowledge and wisdom of the Indian people and refers to their king as the "king of wisdom." He writes about the three Indian astronomical systems—the Sindhind, the Aryabhatiya, and the Khandakhadyaka—then adds, "We have received correct information only about the Sindhind system." He marvels about the remarkable structure of Ḥisāb al-Ghubār (arithmetic of dust or dust-board arithmetic), but is able to cite the name of only one Indian scholar, "Kanka al-Hindi." One may deduce from the text that most of Ṣāʿid's information about India came to him through Baghdad, specifically from Abū Maʿshar's book *al-ʾUlūf*.

The information in the section devoted to the Persians is of little interest; Ṣāʿid mentions without elaborating their contribution to theology and astronomy as it was known to the Arabs of the eleventh century, extracting his information from *Kitāb al-ʾUlūf* and

Andalusia during the time of Ṣāʿid (adapted from Joseph F. O'Callaghan: *A History of Medieval Spain.* Copyright © 1975 by Cornell University).

Kitāb al-ʾIklil. Here Ṣāʿid neglects to mention the many Persian scholars who flourished during the Abbāsid period. Noted among them, we have the illustrious ibn Sīnā (Avicenna), (A.D. 980–1037/A.H. 369–428), and the famous algebraist ʿUmar al-Khayyam (c. A.D. 1020–1110/A.H. 410–503). Another notable omission is Abū al-Ḥasan ibn Abū al-Rijal, of al-Qayrawan, the famous astrologer-secretary of al-Muʿizz ibn Bādīs (r. A.D. 1016–1062/A.H. 406–454).[21]

In chapter 7, Ṣāʿid presents with some precision the contribution of the Chaldeans to the science of astronomy and optics. He enumerates authors and book titles and points out scientific links among Chaldea, Egypt, and Greece. This is important in view of the fact that early Greek scientists established a tradition of visiting Babylon to learn from its magi (*DSB*, Supplement, XV, 672). Understandably, there is no mention of the contributions of the early Babylonian Empire.

The work of the Greeks is well presented in chapter 8; the author gives the names of philosophers, mathematicians, astronomers, and physicians and critically discusses their works. He mentions the seven Greek schools of philosophy and points out the importance of the school of Pythagoras and that of Plato and Aristotle. Ṣāʿid states that Empedocles, Pythagoras, Socrates, Plato, and Aristotle are the leaders among the Greek philosophers, but has high regard for all Greek scholars; "They served humanity with their labor and guided it with their lights." He chides Muḥammad ibn Zakariyā al-Razi for being a detractor of Aristotle. This section is a good concise summary of the Greek scholarly contributions and an indication that by the eleventh century most of the Greek work had been translated into Arabic.

In chapter 9, Ṣāʿid writes about the fifth nation to have cultivated science, the Romans. This is a letdown from the previous chapter. After presenting the history of the development of the Roman Empire, he enumerates the non-Muslim scholars who worked during the Abbāsid period and ignores all Latin contributors.

The following chapter begins with some ancient legends about the land of the Nile and strange animals. Then the author rapidly names the scientists of Alexandria and apologetically notes: "I do not know the definite date of any of the individual scientists of Alexandria . . . or any additional specific information about them." Supplemental information about science in ancient Egypt was deemed appropriate.

A large portion of the manuscript is reserved for the Arabs. In chapter 11, Ṣāʿid writes in general terms about the history, religions, and social customs of the Arabs and recalls some old verses to describe their conditions prior to the advent of Islam.

In chapter 12, the author discusses in some detail the Arab contributions, during the Abbāsid dynasty, to astronomy, mathematics, philosophy, medicine, geometry, and genetics. He praises the role played by Caliph al-Maʾmūn: "his noble soul craved the understanding of wisdom." The giants of the Abbāsid period are also singled out: al-Razi, "the unparalleled physician of the Muslims"; al-Fārābī, "truly, the philosopher of the Muslims"; al-Kindi, "the philosopher of the Arabs and the son of one of their kings"; al-Khuwarizmi, "the mathematician of the Arabs"; and Mūsā ibn Shākir, "one of the best-known of al-Maʾmūn's astronomers."

Chapter 13 is a unique treatment of the scientific developments in al-Andalus. After a precise description of the geography of the country, Ṣāʿid remarks on the influence of the Umayyad caliphs, ʿAbd al-Raḥmān III and al-Ḥakam II, on the manifestation of science in al-Andalus. He laments the eclipse of scientific activities during the reign of al-Ḥājib Abū ʿĀmir, who ordered the destruction of the Umayyad library. This dark period was short-lived and was followed by rapid progress in all the fields of knowledge. With encyclopedic knowledge, Ṣāʿid writes about the scholars who lived in Muslim Spain between the eighth and eleventh centuries, making this section an invaluable and precise reference source about the scientific activities in Muslim Spain during that period. Ṣāʿid's broad picture of science in Muslim Spain is correct, and his limitations reflect on the intellectual environment rather than on his shortcomings.

Here Ṣāʿid's support of Aristotelian philosophy represents a trend in Andalusian thoughts that later culminated in the work of ibn Ḥazm and ibn Rushd [Averroës].

The last chapter of the manuscript is presented by Ṣāʿid as if it were an afterthought; he writes with some authority about the Hebrew scholars of al-Andalus to the neglect of all others, except for the very few who prospered during the Abbāsid period: "the Jewish scientists who specialize in Jewish laws are too numerous to count. . . ." But the inclusion of this chapter is a tribute to the important role the Hebrew scholars played in the development of science in Islamic Spain.

Prior to the Arab invasion of Spain, the Romans were the ruling class in Europe. But the Roman contributions to science were limited at best. With the fall of the Roman Empire around A.D. 455/B.H. 172, Europe was beset by the Dark Ages: overrun by barbarians from the North, it was unable to recover until the Arabs invaded Spain in the eighth century.

The scientific renaissance of the Middle Ages came into Europe through the introduction of Arabic and Greek sciences; most of the work of the Greeks entered into Europe via Spain by means of Arabic, not Latin, manuscripts. During that period, Muslim Spain possessed the writers, and the Greek works from which they drew, to transmit knowledge to Western Europe.

For Europe and Western civilization, the contributions of Islamic Spain were invaluable and included practically all fields of knowledge. Until the end of the eighth century, the Arab culture in Spain was mostly derived from the thriving cultural centers of Baghdad. But Islamic Spain began to make its own contributions during the reign of ʿAbd al-Raḥmān III (A.D. 912–961/A.H. 299–350). He imported books, recruited scholars, and built research centers, hospitals, libraries, and institutions for Islamic studies to make Córdoba a scientific center that rivaled Baghdad.

One of the famous scholars to join the court of ʿAbd al-Raḥmān was ʿAbbas ibn Firnas (d. A.D. 887/A.H. 273). He came to Córdoba to teach music, but soon developed an interest in the mechanics of flight, anticipating Leonardo da Vinci by some six hundred years. He constructed a mechanized planetarium in which the planets actually revolved and influenced the development of astronomy in Europe.[22]

Other Andalusians who made significant contributions to the field of astronomy include Muslamah al-Majriṭi (d. A.D. 1007/A.H. 397 in Córdoba). He was a prolific author with a profound knowledge of both mathematics and astronomy. There was also al-Bitruji [Alpetragius] (fl. A.D. 1190/A.H. 586 in Seville). He developed an adequate theory of stellar movement, and his work includes the *Book of Form* that became very popular in the West. There was also al-Zarqāli, one of Ṣāʿid's accomplished students, as mentioned earlier.

The impact of the work of Andalusian astronomers on the West is partly reflected in the many Arabic terms that are still in use, such as Aldebaran [the follower], alidade, almanac, Altair [the flier], azimuth, Betelgeuse [*bayt al-Jawzāʾ*, house of the twins], Deneb [tail], nadir, zenith, and many others.

In the development of astronomy and mathematics and in a broad sense, Greece was indebted to Mesopotamia, Islam to Greece, Iran, and India, and Western Europe to Islam.

Physicians of Islamic Spain made tremendous contributions to Western medicine; outstanding among them was Abū al-Qāsim al-Zahrawī [Abulcasis or Albucasis] (c. A.D. 936–1013/c. A.H. 324–403). He was probably the most famous surgeon of the Middle Ages. He authored *al-Taṣrif,* a book that was later translated into Latin to become the leading medical textbook in European universities. The section on surgery contains illustrations of functional, elegant precision surgical instruments.

". . . ibn Sīnā (Avicenna) (980–1037 A.D.), the illustrious Shaikh, the prince of all learning! His medical encyclopedia—al-Qānūn (the Canon)—which was for centuries a sort of medical bible—was translated into Latin by Gerard of Cremona. Al-Qānūn was large and forbidding and hopelessly out of reach for the majority of physicians, but every one of them knew of it and thought of it as the supreme monument of medical learning and wisdom." *The Canon of Medicine* dominated the European medical field for centuries and was used as a text at the universities of Montpellier and Louvain until A.D. 1650/A.H. 1060.[23]

As a consequence of their work in medicine, the Muslim scientists of Spain became interested in botany. The most famous Andalusian botanist, ibn al-Bayṭar (c. A.D. 1190–1248/c. A.H. 586–646), wrote a compendium of medical plants and arranged them in alphabetic order for the benefit of his readers. He spent most of his life gathering information in Spain and North Africa. Another botanist, ibn al-ʿAwwām (fl. c. A.D. 1170/c. A.H. 565), wrote a treatise that gives precise instructions for the cultivation and use of hundreds of species of plants.

The most famous geographer of that period was ʿAbd Allah ibn Idris al-Sharif al-Idrisi (A.D. 1100–1166/A.H. 493–561), who, after studying in Córdoba, traveled widely. He then settled in Sicily, where he wrote *The Book of Roger,* a geography of the world. He also engraved, on a silver planisphere, a disk-shaped map that was considered the first scientific map of the world and one of the wonders of that period.

Another famous geographer and world traveler was Muḥammad ibn ʿAbd Allah ibn Baṭṭuṭah. He was born in Morocco in A.D. 1304/A.H. 703 and traveled extensively through most of the known world, spending about eight years in India alone. He kept a diary that became a book titled *Riḥlāt* [Travels]. The book achieved lasting greatness and became a rich source for both geographers and historians, especially about fourteenth-century India, the Maldives, southern Russia, and black Africa (*DSB,* I, 516).

The Andalusian philosopher and physician ibn Rushd Abūʾl-Walīd Muḥammad [Averroës or Averrhoës] (A.D. 1126–1198/A.H. 520–594) was an ardent Aristotelian scholar. After being translated into Latin, his books had a very lasting effect on the development of Western philosophy and medical theories. Ibn Rushd built his philosophy on the work of earlier Andalusian scholars, among them Abū Muḥammad ʿAli ibn Aḥmad ibn Ḥazm (d. A.D. 1064/A.H. 456), who had to address the problem generated by the introduction of Greek thoughts into the context of Islam. Ibn Ḥazm was an intellectual giant, with more than four hundred titles to his credit.

By the twelfth century, European pioneers turned beyond the Pyrenees, seeking the key to knowledge. Al-Andalus remained their principal source until the close of the thirteenth century.[24] During that period, Islamic Spain became the bridge through which the scientific accomplishments and the philosophical legacy of the Arabs and the Greeks passed into Europe. And European scholars began a large-scale, practically indiscriminate conversion of Arabo-Greek learning from Arabic into Latin; literally thousands of treatises were translated.

Leaders of this movement include Robert of Chester, Daniel of Morley, Alfred of Sarashel, Adelard of Bath, Roger of Hereford, John of Seville, Plato of Tivoli, Gerard of Cremona, Marc of Toledo, Stephen of Antioch, and others. These were instrumental in introducing scientific materials from Muslim Spain into Latin Europe, signaling the dawn of the European Renaissance.

It is as exciting to look back and rediscover the past as it is to look forward into the future. When Ṣāʿid wrote *Ṭabaqāt al-ʾUmam,* astrology was a highly respected science, ranked with astronomy and medicine. A knowledge of the influence that the stars have on events on earth was extremely important. Although Ṣāʿid distinguishes between astronomers and astrologers, he names many astronomers who were also astrologers. As a rule it was believed that the signs of the zodiac and the planets control the destiny not only of persons but also of nations, determining their physical characteristics as well as their intelligence and other traits; Ṣāʿid recounts why certain races were scientifically productive and superior, while others were sterile and inferior.

Eleventh-century geographers knew that the earth is spherical, but referred to the land mass between Spain and China as the populated world. There was no mention of continents, but the earth was divided into climatic zones. Ṣāʿid adhered to these concepts.

The Arabs always stressed the importance of genealogy; in many instances Ṣāʿid lists scores of ancestral names tracing the scientist back to a known personality, such as a king, a prince, or a companion of the Prophet. In certain instances, the length of the genealogical list appears to the modern reader to verge on the ludicrous.

Something needs to be said about writing Arabic words for English readers; about half of the twenty-eight Arabic letters have no English equivalent and most of them originate deep from within the chest; thus their correct pronunciation requires some practice.

Every European language except English has formally developed a single standard system for transliterating Arabic words. We follow the guidelines used in the *Concise Encyclopedia of Islam.* If a word is known in some other form, we include the commonly accepted English form in brackets to clarify some ambiguities.

Several Arabic words are commonly used. The definite article *al-* corresponds to the English article "the." Ibn means "son of." *Banū* indicates a family or a tribe. *Abū* means "father of" but is often used to form a nickname, rather than in the biological sense. *ʿAbd* means "servant of," as in ʿAbd Allah [Servant of Allah] and ʿAbd al-Raḥmān [Servant of the Merciful].

The post-Renaissance revolution in scholarly methods, the recent excavations in the Middle East, and the decipherment of several ancient scripts have added significantly to our knowledge of early times. What is available to us now far exceeds what was available to Ṣāʿid in the eleventh century, so we include a few notes for added clarification, most often to support a statement in the manuscript.

The dates in the manuscript are given in accordance with the Muslim calendar that was established by Caliph ʿUmar. Its starting date is the day the Prophet entered Yathrib [Medina], some 200 miles north of Makkah, July 16, A.D. 622. This marks the end of the celebrated Hijrah of the Prophet and the beginning of the first Islamic state. The Muslim year is a lunar year and is designated A.H. for Anno Hegirae.

A lunar month contains 29 days, 12 hours, 44 minutes, 2.8 seconds. A lunar year of 12 months is, therefore, made up of 354 days, 8 hours, 48 minutes, 33.6 seconds, or approximately 354 11/30 days. Thus 32 solar years plus 6.5 days are equal to 33 lunar years. To convert dates from the Muslim calendar to the Gregorian calendar, the following relation is adequate:

$$\text{A.D.} = 622 + (32/33)\ \text{A.H.}$$

S. I. S. AND A. K.

In the name of Allah,

the Merciful,

the Compassionate

Ṭabaqāt al-ʾUmam

Chapter 1

The Seven Original Nations[1]

This is Kitāb Ṭabaqāt al-ᵓUmam written
by Ṣāᶜid, Mercy be upon him

The Qāḍi Abū al-Qāsim Ṣāᶜid ibn Aḥmad ibn Ṣāᶜid wrote:[2] it is known that all the people on earth from the East and from the West, from the North and from the South, although they constitute a single group, differ in three distinct traits: behaviors, physical appearances, and languages.

Those interested in the history of nations, in the study of the chronological record of human events, who have searched the *ṭabaqāt* of the centuries claim that all the people of the distant past, before the branching of tribes and the division of languages, formed seven nations.[3]

The first nation was al-Furs [the Persians], who inhabited the center of the populated world. Their territory bordered on the mountains in northern Iraq, reaching ᶜAqabat Ḥulwān [Gulf of Zagros], which includes Māhāt, Karaj [*EI,* IV, 1056], Dinūr, Hamadan, Qumm [*EI,* V, 369], Qashān, and others. These borders extended to the countries of Armenia, al-Bab wa al-Abwab [Darband, *EI,* I, 189], which is situated on the sea of Khazar [*EI,* IV, 1172], and to Azerbaijan, Ṭabaristan, Muqān, Baylaqān, Rān, Shabarān, Talaqān, and Jurjān, and to the land of Khorasan [*EI,* V, 55], such as Nishapur, Murwan, Sarkhas, Hirāt, Khawarizm, Balakh, Bukhara, Samarkand, Farghana, and Shash, and other cities in the land of Khorasan, Sigistan, Kirmān [*EI,* V, 147], Faris, Ahwāz, Isbahān [Isfahān or Ispahān], and other neighboring countries. They had one kingdom, one king, and one language—Farsi [Persian]. Although they spoke slightly different dialects, they were in agreement on the shapes and the number of the letters in their alphabet, and their differences did not affect the various aspects of those languages, such as the Pahlavi and the Dari as well as other languages of the Furs.

The second nation was the Kildaniyūns [Chaldeans]; they are the Sirianiyūns[4] and the Babylonians; those peoples were comprised of the Kanᶜaniyūns [Canaanites], the Assyrians, the Armenians, and the Jaramiqah [Masᶜūdī 1966–1970:118–121], who inhabited Mussil [Mosul], and the Nabaṭiyūns, who inhabited most of Iraq. Their country was also in the center of the populated world and included Iraq, al-Jazirat [the island] located between the Tigris and the Euphrates [rivers] and known as Diar Rabyᶜah, Mūḍar, al-Sham,[5] and the Arabian Peninsula, which included Ḥijāz, Najd, Tahāmah, Ghūr, and Yemen. All of these are located between Zabīd, Sanᶜāʾ, ᶜAden, ᶜUrūḍ, Shaḥr, Ḥaḍramūt, Oman, and other parts of the Arabian countries. All this was one kingdom having one king and one language, Syryāni [Syriac]. This is the ancient language spoken by Adam, Idris, Noah, Ibrahim, Lot, and others. Later on, Syriac branched out into Arabic and Hebrew. The Hebrews, also known as Banū Israel, conquered al-Sham and inhabited it. The Arabs took over Jazirat al-ᶜArab [the Arabian Peninsula], which is also known as the land of Rabyᶜah and Mūḍar, and inhabited all of it. The rest of the Sirianiyūns had retreated into Iraq, where the capital of their kingdom was Kalwaza [Kalwadha, Masᶜūdī 1966–1970:70, 602).[6]

The third nation was comprised of the Greeks, the Romans, the Ifranjah [Franks], the Jalāliqah [Austrians], the Burjāns, the Slavonians, the Russians, the Burghuz, and the Llān as well as other nations living around the Sea of Neiṭosh [a corruption of Puntus, Black Sea], the Lake of Mayṭus [probably the Sea of Azov], and other areas in the Northwest quadrant[7] of the populated earth. They had one kingdom and spoke the same language.

The fourth nation was the Copts; they are the people of Egypt and the people of the South. They are the Sudanese [black people] from Abyssinia, Nubia, the Zinj [black people, most of central Africa], and others. Also the people of the Maghrib [the West] and they are the Berbers and their neighbors to the west bordering on the Sea of Uqiyanus [Atlantic Ocean]. They spoke the same language and had one kingdom.

The fifth nation was the races of Turks, which include the Karluks, the Kimaks, the Taghuzghuz, the Khazars, the Sarirs, the Jilans [*EI,* II, 111], the Khūzāns, the Ṭilsāns, the Kazakhs, and the Burṭās (*EI,* I, 1337]. They spoke one language and had one kingdom.

The sixth nation was India and Sind and neighboring peoples. They had one language and one king.

The seventh nation was China and the neighboring country of ᶜAmūr [between Manchuria and Russia], named after ᶜAmūr, the son of Yafith, son of Noah. Their kingdom was one and their language was one.

These seven nations comprised the entire human race, and they were all *ṣāʾbat* [Sabians],[8] worshiping idols representing celestial objects, including the seven planets and others.[9] Later on, these seven nations were dispersed, their languages branched out, and their religions became different.

Chapter 2

The Two Categories of Nations

We have determined that all these nations, in spite of their differences and the diversities of their convictions, form *ṭabaqatayn.*[1] One *ṭabaqat* has cultivated science, given rise to the art of knowledge, and propagated the various aspects of scientific information; the other *ṭabaqat* did not contribute enough to science to deserve the honor of association or inclusion in the family of scientifically productive nations. Members of this group formulated no useful philosophy and generated no practical idea.

The *ṭabaqat* that cultivated science is comprised of eight nations: the Indians, the Persians, the Chaldeans, the Greeks, the Romans, the people of Egypt, the Arabs, and the Hebrews.

The *ṭabaqat* that showed no interest in science is comprised of all the remaining *ʾumam* [nations][2] that were not previously mentioned. This includes the Chinese, Hajūj and Majūj [Gog and Magog],[3] the Turks, the Burṭās, the Sarirs, the Khazars,[4] the Gilāns, the Ṭilsāns,[5] the Murqāns, the Kazakhs,[6] the Alains,[7] the Slavonians, the Bulgarians, the Russians, the Burjāns, the Berbers, and the various people of Sudan [or black people],[8] including the Ethiopians, the Nubians,[9] the Zinj, and the Ghanaians,[10] as well as others.[11]

Chapter 3

Nations Having No Interest in Science

Of the nations who have shown no interest in science, the most advanced are the Chinese and the Turks. The Chinese are the most numerous of all nations. They have the richest kingdom and occupy the largest territory. They inhabit the eastern sector of the populated world between the equinoctial line and the farthest of the seven regions to the north. They surpassed other nations in industrial technology and graphic arts. They excelled in their endurance while performing arduous labor and in improving their work and perfecting their products.

The Turks are also a nation having a large population and a rich kingdom. They inhabit the region between the eastern regions[1] of Khorasan of the Islamic kingdom, the western regions of China, and the northern parts of India to the end of the populated world to the north. They distinguished themselves by their ability to wage wars and by the construction of arms, and by being the ablest horsemen and tacticians.[2] They have the sharpest eyes when it comes to throwing lances, striking with swords, or shooting arrows.

The rest of this *ṭabaqat,* which showed no interest in science, resembles animals more than human beings. Those among them who live in the extreme North, between the last of the seven regions and the end of the populated world to the north, suffered from being too far from the sun; their air is cold and their skies are cloudy. As a result, their temperament is cool and their behavior is rude. Consequently, their bodies became enormous, their color turned white, and their hair drooped down. They have lost keenness of understanding and sharpness of perception. They were overcome by ignorance and laziness, and infested by fatigue and stupidity. Such are the Slavonians, the Bulgarians, and neighboring peoples.

Also in this category are the people who lived close to the equinoctial line and behind it to the end of the populated world to the south. Because the sun remains close to their heads for long periods, their air and their climate have become hot: they are of hot temperament and fiery behavior. Their color turned black, and their hair turned kinky. As a result, they lost the value of patience and firmness of perception. They were overcome by foolishness and ignorance. These are the people of Sudan who inhabited the far reaches of Ethiopia, Nubia, the Zinj, and others.

The Jalāliqah [Galicians, Austrians], the Berbers, and the rest of the populations of the western sector that belong to this *ṭabaqat* are nations that Allah, may He be glorified, has provided with despotism, ignorance, enmity, and violence. This is in spite of the fact that these peoples did not inhabit the far North or the far South to be punished by severe climates. In fact their domain is close to the temperate zones. The Galicians reside in the western parts of the fifth climatic region and some of the neighboring provinces of the sixth region. The Berbers inhabit the western parts of the second climatic region, and what borders it from the third and fourth regions. But Allah provides generously for whomever He chooses and diverts His grace away from whomever He chooses.

All the other peoples of this *ṭabaqat* that I have not previously included are similar in their ignorance to those already mentioned. Although they have varied classifications and unequal shares [of ignorance], they belong to the same class and may be described by the same words. They have never searched for wisdom or practiced the study of philosophy. Most of them are an urban population, while their servants live in rural areas. No matter where they are, in the East, in the West, in the South, or in the North, they are governed by royal decrees and divine laws. The only peoples that reject these humane institutions and live outside these rational laws are a few of the inhabitants of the deserts and the wilderness such as the beggars of Bajah [Boga],[3] the savages of Ghana, the misers of the Zinj, and those resembling them.

Chapter 4

Nations That Cultivated the Sciences

The *ṭabaqat* that cultivated science is comprised of the elites that Allah has chosen from among His creatures. They have focused their attention to achieve the purity of soul that governs the human race and straightens its nature [behavior]. They have disregarded what was attractive to the Chinese and the Turks and other people like them who compete with anger in their souls and pride themselves in their animal [physical] strength. These nations ought to know that they share some of these qualities with animals that in many instances surpass them. For example, in the fine construction and the perfection of shapes, the bees prove their superiority in building the cells for storing their food. And the careful spiders construct the strands of their homes to harmonize with the circles that they intersect.

Other animals have also accomplished extraordinary works and strange activities, so much so that the Arabs use them in their proverbs. For example, they say "more industrious than *surfat* [probably the larvae of the caddis fly]." This is a worm that has the ability to build a square home with small sticks. They also say "more industrious than a *tanūṭ* [finch]." This is a bird whose remarkable ability to construct enables him to build his nest suspended from a tree. As to courage and tenacity, the lion, the tiger, and other wild cats are superior to human beings. And men never pretended otherwise. There are several other traits where animals surpass humans, such as generosity, stinginess, and others. The Arabs use them in their proverbs. They say "more generous than a rooster and stingier than a wolf, more audacious than a lion or a fly, more cunning than a fox or a lizard, more humble than a dog, more potent than a snake, more studious than an ant or a wolf, more negligent than an ostrich, more able to find its way than a homing pigeon, more careful than a crow, stingier than a dog, more persistent than a bug, more cunning than a fox, more patient than an old horse, more timid than a nightingale, more loving than a camel."[1]

Similarly, when it comes to physical strength and keenness of senses, no one can deny that the share of some animals is larger than that of humans. For that reason, the Arabs say in their proverbs "having sight better than that of a hawk" or "a horse," "more robust than a wolf" and "a hyena," "stronger than an ant," because of its ability to drag a seed several times its weight, having "hearing better than that of a monkey" or "a horse in the desert" or "a porcupine," "faster than a stallion." The Arabs have many other similar proverbs dealing with the various characteristics of animals.

For the noble reason and the honorable quality that demonstrate human values, the nobility of human virtues, and the superiority of humans to other creatures, scientists came into being. They have eliminated any similarities between human beings and animals and have elevated the human race above the lions and tigers. These scholars are flambeaux in the darkness. They are the guides to wisdom, the masters of the race, and the elite of the nations. They have understood what God wanted them to do, and they have known the goal assigned to them. May Allah's prayers be upon them. And how desolate the world would have been if they were not in it!

As we have mentioned, the *ṭabaqat* that cultivated the sciences is made up of eight nations.[2] Our intention is to introduce their sciences and draw attention to their scholars. Let us begin to do so in accordance with our concise and condensed style, Allah willing.[3]

Chapter 5
Science in India

The first nation [to have cultivated science] is India. This is a powerful nation having a large population and a rich kingdom [possession]. India is known for the wisdom of its people. Over many centuries, all the kings of the past have recognized the ability of the Indians in all the branches of knowledge.

The kings of China have stated that the kings of the world are five in number and all the people of the world are their subjects. They mentioned the king of China, the king of India, the king of the Turks, the king of the Furs [Persians], and the king of the Romans.[1]

They referred to the king of China as the "king of humans" because the people of China are more obedient to authority and are stronger followers of government policies than all the other peoples of the world. They referred to the king of India as the "king of wisdom" because of the Indians' careful treatment of *ʿulūm* [sciences] and their advancement in all the branches of knowledge. They referred to the king of the Turks as the "king of lions" because of the courage and the ferocity of the Turks. They referred to the king of Persia as the "king of kings" because of the richness, glory, and importance of his kingdom, since Persia had subdued the kings of the center of the populated world, and because it controlled, to the exclusion of other kingdoms, the most fertile of the climatic regions. And they referred to the king of the Romans as the "king of men" because the Romans, of all the peoples, have the most beautiful faces, the best-built bodies, and the most robust physiques.

The Indians, as known to all nations for many centuries, are the metal [essence] of wisdom, the source of fairness and objectivity. They are peoples of sublime pensiveness, universal apologues, and useful and rare inventions. In spite of the fact that their color is in the first stage of blackness, which puts them in the same category as the blacks, Allah, in His glory, did not give them the low characters, the poor manners, or the inferior principles associated with this group and ranked them above a large number of white and brown peoples.

Some astrologers came up with an explanation for this condition; they said that both Saturn and Mercury control the destiny of the

Indian people. Because of the influence of Saturn, their color turned black, while the influence of Mercury provided them with intellectual power and fine spirit. Saturn in partnership with Mercury gave them correctness of reasoning and depth of perception. This is why they enjoy the purity of talent and the power of distinction, making them totally different from the people of Sudan [black people][2] such as the Zinj, the Abyssinians, the Ethiopians, and others. To their credit, the Indians have made great strides in the study of numbers[3] and of geometry. They have acquired immense information and reached the zenith in their knowledge of the movements of the stars [astronomy] and the secrets of the skies [astrology] as well as other mathematical studies. After all that, they have surpassed all the other peoples in their knowledge of medical science and the strengths of various drugs, the characteristics of compounds, and the peculiarities of substances.

Their kings are known for their good moral principles, their wise decisions, and their perfect methods of exercising authority.

As to theology, they are in agreement as to the unity of God, may His power and glory be proclaimed, and they exalt Him above any polytheism, but they are in disagreement about His various manifestations.[4] Some of them are Brahmins and others are Sabians.[5] The Brahmins are few in number and are descendants of noble ancestry. Some of them believe in the creation of the world, while others believe in its eternity. But they are in agreement as to the banning of prophecies, prohibiting the slaughter of animals, and refraining from maltreating them or eating their food.[6] But the Sabians, and they are the overwhelming majority of the Indians, believe in the eternity of the world because it is created by the creator of creators, God Himself, may His power and glory be proclaimed. They revere the stars and represent them in paintings and approach them with offerings, each in accordance with what they have learned about its nature. Thus, they obtain the power of each star and use it in the lower world in accordance with their wishes. They believe in the times of return, in the revolutions of planets and their orbits, in the destruction of all the derivatives of the four elements every time the seven planets meet in the head of the lamb [Aries], and in the recreation of compounds during every cycle.[7] On this matter, they have numerous views and a variety of doctrines, as we have indicated in our book *Articles about the Doctrines and Religions of Peoples.*

Since Indians are far from our country and many kingdoms separate us from them, we have very few of their books. Only a small fraction of their knowledge and a few fragments about their religions have reached us, and we have heard about only a small number of their scholars.[8]

Of the Indians' astronomical systems, the three that are well known are the Sindhind, which means cyclic time, the Ārjbahd [Arjabhar], and the Ārkand [Khandakhadyaka of Brahmagupta].[9] We have received correct information only about the Sindhind system, which was adopted and further developed by a group of Muslim scientists. Among them were Muḥammad ibn Ibrahim al-Fazārī [fl. c. A.D. 760–790],[10] Ḥabash ibn ʿAbd Allah al-Baghdādī [fl. c. A.D. 800], Muḥammad ibn Mūsā al-Khuwarizmi [c. A.D. 800–847], al-Ḥusayn ibn Muḥammad, also known as ibn al-Ādamī [fl. c. A.D. 920], and others.[11] The meaning of Sindhind is *al-dahr al-dāhir* [infinite time or cyclic time].[12] This is the way it was reported by ibn al-Ādamī in his tables of astronomy.

Those who believe in the Sindhind say that all the seven planets and their apogees and perigees meet in the head of Aries once every four thousand thousand thousand years and three hundred thousand thousand years, and twenty thousand thousand solar years.[13] They call this cycle the "period of the universe" because they believe that when all the planets meet in the head of Aries everything found on earth will perish, leaving the lower universe in a state of destruction for a very long time until the planets and their apogees and perigees disperse back to their zodiacs [constellations]. When this takes place the world returns to its original state. The cycle repeats itself indefinitely. The Sindhind advocates give no explanation for this behavior but they claim that for each planet and its apogee and perigee there is an orbit that it completes in a given time, which they call the period of the universe. I have already mentioned that in my book *Written for the Rectification of the Movements of the Stars.*

Those who believe in the Ārjbahd agree with the followers of the Sindhind except for the length of the period of the cycle of the universe. They believe that the planets and their apogees and perigees meet in the head of Aries in one thousandth of the period claimed by the Sindhind, and this is the essence of their doctrine.

The followers of the Ārkand differ from the two previous schools in their description of the movements of the planets and also in the length of the period of the cycle of the universe. I have not been informed of the exact nature of this difference.[14]

What has reached us from the work of the Indians in music is the book known in the Indian language as *Bafir* [Nafir], which means *Thimār al-Ḥikmah* [The Fruits of Wisdom]. It contains the fundamentals of modes and the basics in the construction of melodies.

What has reached us from their works on the improvement of morals and the amelioration of upbringing is the book *Kalīlah wa Dimnah,* which was brought by the Persian Ḥakīm [physician or wise] Burzuwaih from India to Anusharwan ibn Qibād ibn Fayrūz, king of Persia [fl. A.D. 550]. Burzuwaih translated *Kalīlah wa Dimnah* for the king from the Indian language to Farsi. It was later translated from Farsi into Arabic during the Islamic period by ᶜAbd Allah ibn al-Muqaffaᶜ.[15] This is a book of noble purpose and great practical worth.

That which has reached us from their work on numbers is Ḥisāb al-Ghubār [dustboard arithmetic],[16] which was simplified by Abū Jaᶜfar Muḥammad ibn Mūsā al-Khuwarizmi. This method of calculating is the simplest, fastest, and easiest method to understand and use and has a remarkable structure. It is a testimony to the intelligence of the Indians, the clarity of their creativity, and the power of their inventiveness.

That which has reached us from the discoveries of their clear thinking and the marvels of their inventions is [the game of] chess. The Indians have, in the construction of its cells, its double numbers, its symbols and secrets, reached the forefront of knowledge. They have extracted its mysteries from supernatural forces. While the game is being played and its pieces are being maneuvered, the beauty of structure and greatness of harmony appear. It demonstrates the manifestation of high intentions and noble deeds, as it provides various forms of warnings from enemies and points out ruses as well as ways to avoid dangers. And in this there is considerable gain and useful profit.

Of the Indian scientists who are knowledgeable in the shape of the physical universe and in the composition of the celestial spheres and the movement of stars, we have heard of Kanka al-Hindi [the Indian].[17] Abū Maᶜshar Jaᶜfar ibn Muḥammad ibn ᶜUmar al-Balkhi [A.D. 787–886] mentioned in his book *al-ʾUlūf* [the Thousands][18] that this scientist is considered a leader in his knowledge of astronomy by all the Indian scientists of the past. I have received no information as to when or where he lived or anything about his work or his life except what I have just mentioned.

Chapter 6

Science in Persia

The second nation [to have cultivated science] is that of the Persians. These are people of high glory and great nobility. Among the nations, they are the closest to the center [of the populated world]. They have the best climate and the ablest kings. We do not know of any other nation that was sovereign for as long or had as many able kings to unify its people and to defend them against all attackers. The Persian kings fought with the people against their enemies and defended the oppressed against the oppressors. They provided them with proper conditions that suited them well for a long time and without interruption. They lived a great life in perfect harmony. Their later kings learned from those who came before them and the descendants learned from their ancestors.

Persons knowledgeable in the history of nations differ as to the duration of the Persian kingdom, but this is not the topic of discussion here, as we have pointed out such differences in our book *Fi Jawāmiᶜ Akhbār al-ᵓUmam Min al-ᶜArab wa al-ᶜAjam* [Compilation of the Annals of Nations from the Arabs and the Non-Arabs]. The most plausible statement that had been made on this subject is that from the reign of Kayumart [Kayomarth] ibn Amim ibn Lawud ibn Sam ibn Noah, the father of all the Persians and considered by them as Adam, the father of all the people, to the beginning of the reign of Minushahr, the first of the kings of the second *ṭabaqat* [dynasty] of the Persian kings, is about one thousand years. From the reign of Minushahr to the beginning of the reign of Kyqabād ibn Raᶜ, the first of the kings of the third *ṭabaqat* of the kings of Persia, is about two hundred years. And from the reign of Kyqabād to the beginning of the reign of King Sassan of the fourth *ṭabaqat* of the kings of Persia, that is, when Alexander killed Dara ibn Dara [Darius III], the last king of the third *ṭabaqat* of the kings of Persia, is about one thousand years. From the beginning of the reign of al-Ṭawāᵓif[1] to the beginning of the reign of Ardshir ibn Babek al-Sassani, the first of the kings of Banū Sassan [Sassanid Dynasty, r. A.D. 226–651], which is the fifth *ṭabaqat* [dynasty] of the kings of Persia, is 531 years. And from the beginning of the reign of Ardshir ibn Babek until the disappearance of the Persian kingdom from the face of the earth, that is, when Yazdijird ibn Shahriyar was killed during the reign of ʿUthmān ibn ʿAffān [r. A.D. 644–656]—may Allah confer grace upon him—in the year 32 of al-Hijrah [A.D. 653] is 433 years. This adds up to a total of 3,164 years.

We have mentioned the duration of the Persian kingdom, although this is not the object of this book, only to show the greatness of their empire and the grandeur of their power. For this as well as other examples of their nobility, these kings deserve to be called the "kings of kings" as we have already mentioned [see chap. 5].

The magnificent quality of the Persian kings, for which they became famous, is their superior political and administrative ability. This is specially true in the case of the kings of Banū Sassan [the Sassanids]. Among them there were kings that surpassed all others in their nobility and their conduct, their ability to govern and exercise authority as well as their widely recognized fame.

The most noted traits of the Persians are the following: their great interest in the study of medicine and their profound knowledge of the positions of stars and the effect of the stars on the lower world. They had ancient observatories to study the planets and they had devised various systems to describe their motions. Among these systems is the one upon which Abū Maʿshar Jaʿfar ibn Muḥammad ibn ʿUmar al-Balkhi [A.D. 787–886] built his grand history and mentioned that it was the system of the leading scientists of Persia as well as many other countries. It was said that the Persians believe that the duration of the world is one twelve-thousandth of the period of the Sindhind;[2] that is, 360,000 years. To them this period is the time required by the mean positions of the planets to meet in the head of the lamb [Aries], without their apogees or their perigees.

Abū Maʿshar,[3] in support of this doctrine, said that mathematicians from Persia, Babylon, India, China, and most other nations who have knowledge of the science of stars, and especially Kanka al-Hindi [see chap. 5], who is the leader of all Indian scientists for all times, agree that the most correct of all astronomical cycles is the one devised by this group, and they used to call it the "period of the world." All the nations of the past who were interested in astronomy referred to it by this name. But the astronomers of our time call it the "period of the Persians."

The Persians wrote important treatises on astronomical systems. Among them is the book *Ṣuār Darajāt al-Fulk* [Illustrations of the Degrees of the Skies], attributed to Zaradusht [Zoroaster], and the book of *al-Tafsyr* [The Commentary] and the book of *Jāmāseb*,[4] and this is extremely important.

A few historians have mentioned that the ancient Persians were all unified under the religion of Noah—peace be upon him—until Budasaf al-Mushriqi [from the Orient][5] came to Tahmurath, the third of the kings of Persia, and presented him with the religion of Ḥanif and they are the Sabians. He accepted this religion from Budasaf and forced the Persians to practice it. They embraced it for about 1,800 years; that is, until they were all converted to Majus [Magus]. The reason for adopting this new religion is that Zoroaster the Persian appeared during the time of Bistāsif [Vishtaspa][6] some thirty years after he became king, and preached the religion of the Majus, which glorifies fire and other forms of light and teaches that the world is made up of light and darkness and adheres to five basic doctrines that are God the creator, Iblis [Satan], Matter, Space, and Time, as well as other tenets pertaining to the Majus religion.

Bistāsif, after accepting this religion from Zoroaster, fought the Persians and forced all of them to embrace it and renounce the religion of Sabians. In the end, the Persians accepted Zoroaster as a prophet sent to them by Allah—may His glory be proclaimed. They continued to practice this religion and embraced its tenets for 1,300 years, until ʿUmar ibn al-Khaṭṭāb—may Allah confer grace upon him—conquered their kingdom and took over their cities, the centers of their glory, and threw them out of Iraq and the neighboring provinces of Khorasan.

The rest of their kingdom was uprooted by ʿUthmān ibn ʿAffān—may Allah confer grace upon him. During his caliphate, he killed Yazdijird ibn Shahriyar, the last of the Persian kings, in the year 32 of the Hijrah [A.D. 653] and killed a large number of Persians during the wars between them and the Muslims, as on the day [battle] of al-Qādisiyah [A.D. 636], on the day of Jalūlā [A.D. 637], on the day of Nihawand [A.D. 640], and on other days. Some of them accepted Islam while the rest continued to practice the religion of Majus until the present time. They still live like the Jews and the Christians in Iraq, Ahwāz, Persia, Aṣbahān, Khorasan, and other provinces that formed the Persian kingdom before Islam.[7]

Chapter 7

Science of the Chaldeans

The third nation [to have cultivated science] is that of the Chaldeans. This was a nation of ancient heritage and intelligent kings. A few of them were known as the powerful Namrūds [Nimrods—despots].[1] The first Nimrod was ibn Kūsh [Cus], the son of Ham, the builder of the tower who was mentioned by Allah in the Qur'an: "Those before them schemed, and Allah reached for their dwelling through its foundations, so the roof caved in on them and torment came at them from whence they did not suspect."[2] Abū Muḥammad al-Ḥasan ibn Aḥmad ibn Ya^cqūb al-Hamdānī [c. A.D. 893–951], also known as ibn Dhi al-Damynah, the author of the book *Sarā'ir al-Ḥikmah* [Secrets of Wisdom] and the book *al-'Iklil* [The Crown] as well as others, stated that the height of the roof of the tower, as indicated by other scientists, was 5,000 *dirā^c* [arm-lengths][3] and its width was 1,500 *dirā^c*. The Babylonians believed that this Nimrod who constructed the tower was the first king on earth after the Flood. Among the kings of Babylon there was Nimrod Ibrahim—peace be upon him. He is Nimrod, son of Cana^can ibn Sinḥārīb, son of Nimrod the Great, the owner and constructor of the tower. Among them, there was also Bught Nasir [Nebuchadnezzar] ibn Fayrūzādān ibn Sinḥārīb, one of the sons of the Nimrod the Lesser, son of Cana^can, who invaded Banū Israel, killed many of their people, and took the others captive. He also invaded Egypt and threatened many other countries. Nebuchadnezzar reigned over Babylon and all the Chaldean Empire until his defeat at the hands of the Persians, who conquered his kingdom and killed many of its people. Thus their history was lost and their traces were erased.[4]

Among the Chaldeans, there were many great scholars and well-established savants who contributed generously to all the branches of human knowledge, especially mathematics and theology. They had particular interest in the observation of planets and carefully searched through the secrets of the skies. They had well-established knowledge in the nature of the stars and their influence. They knew of the properties and strength of the various [chemical] compounds. They instructed the people of the western section of the populated world in the construction of temples and in their use to extract the power of the planets and expose their nature. They knew how to project their lights by various suitable means [principles] and practical applications. Thus the Chaldeans performed marvelous work and attained noble and extraordinary results in the production of talismans as well as other forms of magic.

The best-known and the noblest of their scholars is Hermes of Babylon, who was a contemporary of the Greek philosopher Socrates. He was mentioned by Abū Maʿshar Jaʿfar ibn Muḥammad ibn ʿUmar al-Balkhi in the book *al-ʾUlūf* [The Thousands] as being the one who corrected many of the writings of previous authors in the science of astronomy as well as other forms of philosophy. He authored many books on a variety of topics. Abū Maʿshar said that Hermes were numerous. Their first was the one who came before the Flood, whom the Hebrews refer to as the prophet Khanukh or Akhnukh [Eʾnoch];[5] this is Idris—peace be upon him. After the Flood there were several Hermes noted for their knowledge and wisdom. Two of them stand out: the first was the Babylonian who was mentioned earlier and the second was a student of Fithagoras [Pythagoras, c. 582–500 B.C.], the sage, and he inhabited Egypt.[6]

We have obtained from the doctrines of Hermes the Babylonian that which proves his greatness as a scientist. Among his works are *Maṭāriḥ Shuʿaʿāt al-Kawākib* [Projections of the Light of the Stars] and *Taswiat Buūt al-Falak* [Equalization of Astronomical Dwellings]. As an example of his books on astronomy, we cite *Kitāb al-Ṭūl* [The Book of Longitude], *Kitāb al-ʿArḍ* [The Book of Latitude], and his book *Qaḍib al-Dhahab* [Stick of Gold].

Among the Chaldean scientists who came after Hermes, we have Burkhus [Hipparchus, c. 190–125 B.C.],[7] who wrote the book *Asrār al-Nujūm fi Maʿrifat al-Millal wa al-Duwal wa al-Malāḥim* [Secrets of the Stars in the Knowledge of the Doctrines, the Countries, and the Epics], and Wālys [Valens], the king and author of *Kitāb al-Ṣuwar* [Book of Illustrations] and the book *al-Baraydaj*,[8] which deals with genetic codes and drifts. Also among them was Iṣṭifan [Stephenus], the Babylonian who authored a famous book on the behavior of stars. He lived at the time of the prophet Shuʿayb [Jethro]—peace be upon him.

We have not received detailed information or complete segments from the Babylonians that illustrate the motion of the stars or describe the features of the skies. We do not have, from their views on this subject or from their observations, any direct knowledge,[9] except what was transmitted to us through the writings of Claudius Ptolemy, the Greek, in his book *al-Majisṭi* [Almagest].[10] Ptolemy felt the need for this information to correct the trepidant motion of the stars because he could not trust the validity of the observations made by his compatriots, the Greeks.[11]

Chapter 8
Science in Greece

The fourth nation [to have cultivated science] is the Greek. Among the nations, it has been considered the one with greater power and a widely recognized reputation. Its great kings had earned the respect of all the people of neighboring countries. Among them was Alexander, son of Philippus [Philip] of Macedon, known by the name Alexander Dhu al-Qarnayn [having two horns],[1] who conquered Dara [Darius] ibn Dara, king of Persia, in his own territory. He brought down his throne, dismembered his kingdom, and dispersed his people. Then he traveled beyond, intending to meet the kings of the East, such as those of India, Turkey, and China. He defeated some of them, and all of them accepted his authority. They met him with precious gifts and evaded him with excessive tributes. He kept wandering in the farthest provinces of India and along the borders of China and the rest of the East, until all the kings of the earth without exception submitted to his authority and were humiliated by his glory. They accepted him as king of the world and master of the earth.

After him came a group of Greek kings known as the Baṭālisah, whose singular is Baṭlymus [Ptolemy],[2] and by them many kingdoms were subdued and many heads were lowered. They continued to reign until they were defeated by the Romans. Their sovereignty disappeared from the face of the earth, and their kingdom was united with the kingdom of Rome and the two became one kingdom, the Roman kingdom. This is similar to what the Persians did to the kingdom of Babylon when they conquered it and declared the two kingdoms one Persian kingdom.

The country of the Greeks is situated in the northwestern quarter of the earth. To the south it borders on the Roman Sea [Mediterranean] and parts of al-Sham and Mesopotamia, and to the north on the countries of the Alains [Alans or Alanni] and the neighboring Norse countries, and to the west on the country of the Romans, whose capital is Rome, and to the east on the Armenian provinces and al-Bab wa al-Abwab as well as the gulf that connects the Roman Sea and the north Sea of Neiṭosh [Black Sea] and cuts through the middle of the Greek country, leaving a large portion of it to the east and a small portion to the west. The language of the Greeks is called al-

Ighriqiyah [Hellenic or Greek] and it is one of the richest and noblest of all languages.

Most of the Greeks were Sabians who venerated the heavenly bodies and persisted in their worship of idols. Their scientists were called *falāsifah* [philosophers], whose singular is *faylasūf*. This word in the Greek language means [lover of wisdom]. The Greek philosophers are the highest *ṭabaqat* [class] of people and the most respected among the people of knowledge; this is because of the true care that they have demonstrated in cultivating all the branches of knowledge, including mathematical sciences, logic, natural philosophy, and theology, as well as the political sciences that deal with the home and family and the society as a whole.

The greatest of the Greek philosophers are five: historically the first one is Bindaqlis [Empedocles, c. 490–430 B.C.], then Pythagoras [c. 582–500 B.C.], Suqrāṭ [Socrates, c. 469–399 B.C.], Aflāṭūn [Plato, c. 427–347 B.C.], and Arisṭotālīs [Aristotle, c. 384–322 B.C.], the son of Nicomachus [Nicomack]. There is a general agreement that those five are the ones who deserve to be called philosophers of Greece.[3]

As reported by those knowledgeable in the history of nations, Empedocles lived during the time of David—may peace be upon him—and studied philosophy with Lukman the sage,[4] in al-Sham. Then he traveled to Greece, where he expressed opinions about the creation of the world that appeared to taint the notion of al-Miyᶜad [resurrection?]. For that reason some have deserted him, while a religious faction known as *al-bāṭiniyah*[5] subscribed to his philosophy and claimed that he portrays complex and mysterious ideas that could be apprehended only rarely. Al-Bāṭinī, Muḥammad ibn ᶜAbd Allah ibn Musrah al-Jabalī [d. c. A.D. 932], of the city of Córdoba was fond of his philosophy and persistent in its study. Empedocles was the first to attempt a unification of the attributes and concepts of God and to declare that it all leads to the same thing [Being], and that the descriptions of science, excellence, and power do not carry different meanings when attributed to these various names. In reality, He is the one [unique], indivisible in any form and different from all other forms. Other universal entities are subject to subdivisions either in their parts or in their meanings or in the concepts that they represent. But only the essence of the creator is above all this. Abū al-Hudhayl Muḥammad ibn al-Hudhayl al-ᶜAllaf al-Baṣrī did subscribe to this doctrine of attributes.[6]

Pythagoras came a long time after Empedocles.[7] He studied philosophy in Egypt under the disciples of Solomon, son of David. These disciples left Palestine to live in Egypt. Prior to that, he studied geometry under the tutelage of Egyptian masters. Later on, he returned to Greece and introduced geometry, physics, and theology into that country. Because of his sagacity, he discovered music and the composition of melodies, which he later submitted to numerical analysis and the ratios of numbers. He pretended that he received this information through the light of prophecy. In describing the harmony of the world and its compositions according to the properties and classification of numbers, he presented marvelous allegories and sublime ideas. In discussing future life [life after death], his ideas were similar to those of Empedocles: above the physical world, there exists a spiritual world, a world of light, whose beauty and magnificence may not be apprehended by intelligence alone: the pure soul longs for and aspires to it. Every man who ameliorates himself by ridding himself of vanity, arrogance, hypocrisy, envy, and all other defects born out of desire and lust becomes worthy of association with the spiritual world and of understanding the realities of the divine wisdom. The pleasures of the soul come to him like the tide, and like a beautiful melody it hits his ears. He has no further need to search for it. Pythagoras is the author of valuable treatises on arithmetic, music, and so forth.

Socrates was one of the students of Pythagoras [Socrates, though he lived in a different time, was a Pythagorean philosopher]. In the study of philosophy, he specialized in metaphysical sciences and disdained worldly pleasures. He rejected and publicly proclaimed his disagreement with the Greeks who worshiped idols and faced their leaders with arguments and proofs. Those leaders decided to rouse the masses against him and to force the king to put him to death. To appease them, the king threw him in jail, and later on, to evade their indignation, he had him poisoned [399 B.C.], after having many well-known discussions with him.[8]

Socrates is the author of sublime counsels, remarkable institutions, and famous sayings. His opinions on the divine attributes were similar to those of Pythagoras and Empedocles. But on the subject of future life, his ideas rested on weak foundations, far from sound philosophy and true doctrines.

Plato, like Socrates, adopted the philosophy of Pythagoras, but did not become known for his wisdom until after the death of Socrates. He was of noble origin; his family was known for its many scholars. He was versed in all the branches of philosophy and wrote many famous books on a variety of subjects and taught his philosophy to many students. As he lectured his students while walking, they became known as the Peripatetics [pedestrians]. Toward the end of his days, he delegated the authority of teaching to the ablest among his students and companions and isolated himself from other people and devoted his life to the worship of his God.

{It is said that he was once asked: "What is love?" And he replied: "Love is an agitation of the soul having no purpose or reflection."}[9]

Among his famous works is the book *Fādin* [Phaedo], on the

spirit, the book *Civil Politics* [The Republic], and *Ṭymāwūs al-Rūḥāni* [Timaeus of the Spirit], which deals with the composition of the three intellectual worlds: the world of the divine, the world of rational reasoning, and the world of the spirit, and his book *Timaeus of Physics,* in which he discusses the composition of the physical universe: these two books were written to one of his disciples, named Timaeus.

Aristotle is the son of Nicomachus al-Gehrāshnī [Nicomack of Gerasha], the Pythagorean. Abū al-Ḥasan ʿAli ibn al-Ḥusayn al-Masʿūdī[10] said that Nicomack means "conqueror of the enemy" and Aristotle means "of complete virtue." Nicomack adopted the Pythagorean doctrine and wrote several mathematical treatises. His son Aristotle was a student of Plato, and it is said that he remained with him for some twenty years. Plato preferred him over all other students and named him the "Sage" or the "Brain."

Aristotle is the last and the best known of the Greek philosophers. He was the last of their sages and the most eminent among their scholars. He was the first to separate the art of proof from other forms of dialogues and to provide it with its syllogistic type of argument and turned it into an instrument of theoretical interpretation. For that reason, he was given the name "Logician."

Aristotle wrote on all the subjects pertaining to sciences and philosophy. He wrote great books of a general nature as well as specialized treatises. Each one of his short treatises dealt with a single subject. On the other hand, his general books are mnemonic in nature; they present records of his teachings and philosophies. He wrote some seventy such books. Most of his writings were didactics, written for Ofarus and designed to teach three forms of knowledge: the first is the science of philosophy, the second is the application of philosophy, and third is the instruments used in the science of philosophy as well as other sciences.

Some of his books that deal with the science of philosophy treat mathematical sciences; others treat the physical sciences; and those of the third group are on theology.

1. Among his books on mathematical sciences, there is a book on *Manāẓir* [Optics], a second on *al-Khuṭūṭ* [Lines], and a third on *al-Ḥiyal* [Mechanics].

2. Among his books on physical sciences there are those that provide information common to all objects, while others provide information particular to every unique object. In the first category, there is the book known as *Samīʿ al-Kiyān* [Physics]. This book introduces the number of principles that govern all physical objects and all the objects that obey these principles, as well as the exceptions. The principles are the element and the form. But those that are analogous to the principles are not exact but approximate principles

and are of no significance [nonexisting]. The objects of these principles are time and space, but the exceptions are the vacuum [void] and what is infinite.

His principles, which deal with information particular to every unique object, concern themselves with formed [created] objects, while other parts deal with objects without forms [not created]. The first two chapters of his book *al-Samāᶜ wa al-ᶜĀlam* [The Sky and the World/On the Heavens] discuss objects having no forms. Part of the discussion on "formed" objects is general, while the other is specific. The general treats the transformation and motion of objects. Transformation is discussed in his book *al-Kawn wa al-Fasād* [On Generation and Corruption], while motion is treated in the last two chapters of his book *On the Heavens.* The specific treats the elements and the compounds; the elements are discussed in his book *al-Āthār al-ᶜAlawiyah* [Meteorology]; either the compounds are described in their entirety or their various segments are discussed. The description of the entire entities may be found in *Kitāb al-Ḥayawān* [The Book on Animals] and in *Kitāb al-Nabāt* [The Book on Plants]. The description of segments may be found in *Kitāb al-Nafs* [The Book of Spirit] and in the book *al-Ḥiss wa al-Maḥsus* [On Sense and the Sensible] and the book *al-Ṣiḥat wa al-Ṣuqm* [On Health and Sickness] and the book *al-Shabāb wa al-Haram* [On Youth and Old Age].

3. His work on theology includes the thirteen treatises found in his book *Fi ma Baᶜd al-Ṭabyᶜah* [On Metaphysics]. His books on practical philosophy deal either with *Islaḥ al-Nafs* [Ethics] or with *al-Siyāsah* [Politics]. His books on ethics are his *Grand Book* and his *Small Book* [Nicomachean Ethics], both written to his son, and his third book, entitled *Ūdimya* [The Ethics of Udem]. His work on politics deals with the management of either cities or families.

His work on the instruments used in natural philosophy is found in his eight books on logic. To the best of our knowledge, no one wrote or collected information on this subject before him. This was mentioned by Aristotle toward the end of his sixth book, which is the book *Sofistica* [On Sophistical Refutations], where he said:

> On practical logic and the establishment of syllogism, we were unable to find any previous work upon which to build. We have achieved what we did with hard work and sustained effort. Although we originated and invented it, we have built it on a solid foundation, without omitting anything that should be included as was done in previous scientific discoveries. It is complete, solidly constructed on good foundations, with known goals and clear results. We have introduced it with clear and well-established arguments. We hope that anyone who investi-

gates the subject after us will forgive any error that he might detect, and his gratitude and greatness will make him consider all the work needed for its preparation. He who does everything he can is not to be blamed.

Aristotle was the teacher of Alexander, son of Philippus, son of Alexander of Macedonia. In accordance with his teachings, the king governed his subjects and administered his kingdom until polytheism disappeared from Greece, virtue appeared, and justice reigned. Aristotle authored many sublime messages to the king, in one of which he encourages him to carry the war to Darius, son of Darius, king of Persia; another one is an answer to a letter that Alexander wrote him from India in which he describes what he saw in the house of gold in the highlands of India. This is the house that contains al-Baddah [the Buddha]; these are idols represented with divine precious stones. In his reply, Aristotle advised him to turn away from material things and aspire to eternal happiness.[11]

Those five men are the leaders among the Greek philosophers. There were other well-known Greek philosophers, such as Thales al-Malṭi [c. 640–546 B.C.], a *ṣāḥib* [associate or companion] of Pythagoras; and Dhumakṭraṭus [Democritus, c. 460–370 B.C.], who believed that objects may be subdivided into small parts [atoms] that can be divided no further. He has written several treatises on this subject. There was also Anaxagoras [c. 500–428 B.C.] as well as others who came before Aristotle or were his contemporaries.

After Aristotle came a group who followed in his footsteps and explained his works. The most famous of this group are Themistios, Alexander of Aphrodisy,[12] and Farfūriūs [Porphyry]:[13] those three are the most knowledgeable in Aristotle's writings and the ablest in the study of philosophy after him.

Of the Greek philosophers who came later on, and who lived during the Islamic period and during the period of the Kingdom of Banū ʿAbbās [the Abbāsid caliphate] and who were contemporaries to Yaʿqūb ibn Isḥaq al-Kindi [d. c. A.D. 870],[14] we mention Qusṭā ibn Luka al-Baʿalbeki [of Baalbek] al-Shami,[15] who was famous for his profound work on numbers, geometry, astronomy, logic, and physical sciences; he was also a very able medical practitioner. He wrote concise but remarkable books, of which his book *al-Madkhal ila al-Handasa* [Introduction to Geometry], written in the form of questions and answers, is unparalleled. He also wrote the book *al-Madkhal ila ʿIlm Hayʾat al-Aflāk wa Ḥarakāt al-Nujūm* [An Introduction to the Science of the Appearance of the Skies and the Motions of the Stars] and the book *al-Farq Bayn al-Ḥayawān al-Nāṭiq wa al-Ṣāmit* [The Difference between the Animal That Speaks and the One That Does Not] and the book *al-Farq Bayn al-Nafs wa al-Ruḥ*

[The Difference between the Spirit and the Soul] and the book *Nisbat al-Akhlāṭ* [Ratio of Mixtures][16] and the book *Ghalabat al-Dam* [Supremacy of Blood], and well as other books.

The Greek scholars who specialized in certain aspects of the science of philosophy and studied only one of its subjects are numerous. One of those who specialized in physical sciences and medicine is Abkrat [Hippocrates, c. 460–377 B.C.],[17] the leader of the physical scientists of his time. He lived about one hundred years before the time of Alexander [356–323 B.C.]. He wrote concise and noble books on medicine, of which we mention his book *al-Fuṣūl* [The Seasons] and his book *Taqdimat al-Maʿrifat* [Introduction to Knowledge] and his book *Māʾ al-Shaʿyr* [Water of Barley] and his book *al-Janyn* [The Fetus], and so forth.[18]

Also in this group, we cite Julianus [Galen, c. A.D. 130–200],[19] who lived in the city of Burghmus [Pergamum] in the land of the Greeks. He was the leader of the physicians and the head of the physicists of his time. He is the author of great books on practical medicine and on physical sciences and of *ʿUlūm al-Burhān* [The Science of Proof]. Galen introduced the titles of his books in a large index in which he mentioned the order in which they should be read and the method of studying them. His books numbered more than a hundred. Abū al-Ḥasan ʿAli ibn al-Ḥusayn al-Masʿūdī said that Galen lived about A.D. 200, or about six hundred years after Hippocrates, and a little more than five hundred years after Alexander. I know of nobody except Aristotle who is more knowledgeable in the physical sciences than these two scholars: I mean Hippocrates and Galen.

Among the "naturalists," besides these two, we have Asqlibiyādhus [Asclepiades], Arisṭarkhus [Aristarchus of Samos, c. A.D. 310–230], Lukus [Luke], and Bulus [Paul], who contributed to the physical sciences. But most of their work was of inferior quality and far from being true. Aristotle and Galen pointed that out and demonstrated their errors with solid arguments and conclusive proofs.

Among the Greek mathematicians, we have Ablūniūs the Carpenter [Apollonius of Perga, fl. c. 200 B.C.],[20] who wrote the book on *Makhrūṭāt* [Conics], which discusses bent lines that are neither straight lines nor arc segments; and Iqlidus al-Ṣūri [Euclid of Tyre, fl. c. 300 B.C.],[21] the author of the famous introduction to geometry known by the title *al-Arkān* [The Foundations]. He is also the author of the book *al-Maʿrūḍāt* [Prisms], the book *al-Manāẓir* [Optics], the book *Taʾlyf al-Luḥūn* [Composition of Harmonies], and others.

In one of his books, Abū Yūsuf Yaʿqūb ibn Isḥaq al-Kindi mentioned that one of the Greek kings found in his library two books attributed to Apollonius the Carpenter in which he discussed the characteristics of five solids whose circumference cannot be exceeded by that of a sphere. The king searched for someone to help him understand the two books, but Euclid was the only one who could. He was the best geometer of his time. He simplified the two books and explained to the king what Apollonius meant, then added some diagrams to help him understand these five solids. This resulted in thirteen articles attributed to Euclid, who later added two articles on the ratios of solids and their mutual relations. None of this was mentioned by Apollonius.

Among the Greek scientists, there is also Arshmydis [Archimedes, c. 287–212 B.C.],[22] who authored the book *al-Musabaʿ fi al-Dāʾirat* [Heptagon inside the Circle], the book *Masāḥat al-Dāʾirat* [Area of the Circle], and the book *al-Kura wa al-Isṭiwānat wa al-Makhrūṭ* [The Sphere and the Disk and the Cone].

There is also Faṭūn, who lived toward the end of the Greek Empire. He was a mathematician and a surveyor and wrote well-known books on both subjects.

We should also mention Sinbliqius [Simplicius, c. A.D. 500–533],[23] who came after Euclid, and Kharmidus [Charmides] and Abū Sindrinus [Posindrinus]. There is also Timalaus, known for his astronomical observations, which were mentioned by Ptolemy in one of his books. Ptolemy also indicated that Timalaus lived 420 years before him. There are also Milāus [Menelaus, Greek geometer, 1st century A.D.] and Tāwadsiūs [Theodosius, fl. c. 150 B.C.], author of *al-ʾUkar* [Spherics]. Also, among them, there are Myṭon [Meton, fl. c. second half of fifth century B.C.] and Afṭimon [Euctemon, fl. fifth century B.C.], the two astronomers who lived in Alexandria, Egypt, some 571 years before Ptolemy. Also, we have Abarkhus [Hipparchus, c. 190–125 B.C.],[24] the celebrated astronomer, whose observations and research were remarkable. He lived some 300 years after Meton and Euctemon.

There is also Baṭlymus al-Qaluzi [Claudius Ptolemy, A.D. 85–165],[25] the author of the book *al-Majisṭi* [Almagest], the book *al-Manāẓir* [Optics], and the book *al-Maqālāt al-Arbaʿ* [The Four Articles, Tetrabiblios] on the study of astronomy. He also authored the book *al-Musiqa* [Music], the book *al-Anwāʾ* [Sea Storms] or *al-Anwār* [Lights], and also the book *al-Qānūn* [Canon], which he extracted from *Almagest.* He was a contemporary of Idriyanūs [Hadrian, c. A.D. 76–138] and Anṭonin [Antonius, A.D. 86–161], the two Roman emperors; that is, he lived some 280 years after Hipparchus.

Many people who claim knowledge of the history of nations include Claudius Ptolemy with the Greek Ptolemies who reigned after Alexander. This is clearly an error because Ptolemy mentions in his book *Almagest,* specifically in the third section of the third book where he discusses all the motions of the sun, its observations, and all its variations, that he observed the fall equinox during the nineteenth year of the reign of Hadrian. This means that from the first year of the reign of Nebuchadnezzar [c. 604–561 B.C.] to the time of this fall equinox, there were 879 years, 36 days, and 6 hours. Subdividing this period, he stated that from the first year of the reign of Nebuchadnezzar until the death of Alexander of Macedon, the grandfather of Alexander Dhu al-Qarnayn, there are 424 Egyptian years; and from the death of Alexander to the time of King Augustus [63 B.C.–A.D. 14], the first of the Roman emperors, there are 294 years, and from the first year in the reign of Augustus to the observation of the fall equinox, there are 161 years, 66 days, and 2 hours. Thus Ptolemy demonstrated in a clear and concise fashion that from the time of Augustus to his own time there are 161 years. Those knowledgeable in the history of nations and the annals of past generations have agreed that this Augustus is the Roman emperor who defeated Qlūbaṭra [Cleopatra, 69–30 B.C.], the last of the Ptolemaic Greek rulers, and usurped her kingdom. With this defeat, the Greek kingdom disappeared from the face of the earth. This is enough to clarify the error of those who said that he [Claudius Ptolemy] was one of the Ptolemaic kings. Allah willing.

With this Ptolemy, the science of astronomy and the knowledge of the secrets of the skies reached perfection. He collected all the fragments of this science that were obtained by the Greeks, the Romans, and the rest of the people who lived in the western region of the earth. He organized its parts and clarified its obscurities; and I do not know of anybody after him who even tried to author a book that resembles his book *Almagest* or anybody who tried to criticize it, although some scientists explained some of its parts and clarified some of its contents; such as al-Faḍl ibn Ḥātim al-Nayryzī [from Nayryz, Persia, d. A.D. 922], while others abbreviated it and rendered it more accessible, such as Muḥammad ibn Jābir al-Battānī [d. A.D. 900]. Understanding his book and the arrangement of its parts was the goal of the scientists who came after him; they worked and competed to attain this goal. I do not know of any book dedicated to a given scientific field, whether old or new, that is more complete in its treatment of the field than the following three books: the first is the book *Almagest,* on the science of astronomy and the motion of the stars, the second is Aristotle's book on the study of logic, and the third is the book of Sibawayh al-Baṣrī [from Baṣra] on the study of *al-Naḥū al-ᶜArabi* [Arabic Grammar and Language]. Each one of these three books contains all the essentials and the secondaries of the subjects they treat. What is omitted has no importance whatsoever. And Allah alone is endowed with complete knowledge and perfection. There is no God but Him.

Those were the greatest and the most famous among the Greeks. They served humanity with their labor and guided it with their lights. The Greeks have in addition several philosophers and scholars whose wisdoms and chronicles were compiled by other authors.

It was stated by Ḥunayn ibn Isḥaq al-Turjumān [the translator] and Abū Naṣr Muḥammad ibn Naṣr al-Fārābī al-Manṭiqī [the logician, c. A.D. 870–950] as well as other philosophers that Greeks formed seven schools of philosophy whose names were derived from seven sources: the first is the name of the man who taught the philosophy, the second is the name of the city where the philosophy originated, the third is the name of the place where it was taught, the fourth is from the relations that it inspired, the fifth is from the principles arising from the philosophy, the sixth is from the ideas observed in the goal taught by the philosophy, and the seventh is in the acts that appear during the teaching of that philosophy.

The school whose name is derived from the name of the man who taught the philosophy is that of Pythagoras. The school whose name is derived from the name of the philosopher's hometown is that of Arisṭifūs [Aristippus][26] of Qūrinā [Cyrene in northeast Libya]. The school whose name is derived from the name of the place where the philosophy was taught is the sect of Karsifus [Chrysippus], known as the school of the Stoic [portico or porch][27] because they taught under the portico of a temple in the city of Athens. The school whose name is derived from the conduct and manners of its people is the sect of Diūjānus [Diogenes], known as the Kilāb [Dogs or Cynics]. They were known by that name because they opposed the doctrines imposed on the people in the cities, such as the love of their relatives and the hate of all others. These are the manners found in dogs. The school whose name is derived from the ideas that its disciples derived from the philosophy is the sect of Fūrūn [Pyrrhon]. They were known as people of pleasure, because they believed that the ultimate goal derived from studying philosophy is the pleasure of learning it. The school whose name is derived from the actions of its followers is the sect of Plato and the sect of Aristotle; they were known as Peripatetic or pedestrian because Plato and Aristotle used to teach their students while walking, so that the body exercises when the spirit does. These are the *ṭabaqāt* of the Greek philosophy. The two most important are the school of Pythagoras and that of Plato/Aristotle. These two are the foundations and the pillars of philosophy.

The early Greek philosophers specialized in the study of natural philosophy put forth by Pythagoras and Thales al-Malṭi as well as most of the Sabians whether Greeks or Egyptians. Those who came later, such as Socrates, Plato, Aristotle, and their followers, preferred the philosophy of ethics. Aristotle mentioned that in his book about animals: he said, "During the last one hundred years, that is since the time of Socrates, people have begun to move away from natural philosophy toward the philosophy of moral ethics."

Several thinkers who came later on wrote books about the doctrines of Pythagoras and his followers in which they defended the old natural philosophy. Among those who wrote on this subject, we have Abū Bakr Muḥammad ibn Zakariyā al-Razi [d. c. A.D. 932], who had a great deal of distaste for Aristotle, blaming him for his deviation from the teachings of Plato and other early philosophers. He claimed that Aristotle had corrupted the philosophy and changed many of its basic principles. I believe that al-Razi's distaste for and his criticism of Aristotle are the result of their opposite views, as stated by al-Razi in his book *Fi al-ʿIlm al-ʾIlahi* [On the Science of Theology] and in his book *Fi al-Ṭibb al-Rūḥani* [On Spiritual Medicine] as well as his other works where he demonstrated his preference for the doctrine of dualism in polytheism and for the doctrines of the Brahmans in the repeal of prophecy and the beliefs of the common Sabians in reincarnation. But if Allah provided al-Razi with guidance and assisted him to support the truth, he would have described Aristotle as the authority who understood the ideas of philosophers and examined the doctrines of scholars; he rejected and discarded what is weak and what is bad and chose what is pure and preserved what is best. He adopted from it what is dictated by clear thinking and penetrating views and what is accepted by pure souls.

Thus Aristotle became the leader of the philosophers and one who united all the virtues of the scholars.

> No one can object if Allah
> Assembled the world in one individual.[28]

Chapter 9
Science of the Romans

The fifth nation [to have cultivated science] is the Rūm [Romans].[1] This is a nation of great power and glorious kings; their territory neighbored that of Greece, but their language was completely different from the Greek language: the language of the Greek is al-Ighriqiyah [Hellenic] and that of the Romans is Latin. The Roman territory borders to the south on the Roman Sea [Mediterranean Sea], which extends longitudinally from east to west, between Tangier and al-Sham, and to the north on some of the northern nations such as the Russians, the Bulgarians, and others and on a segment of the great western sea known by the name Uqiyanus [Ocean],[2] and to the east on the confines of the Greek territory, and to the west, at the extreme end of Andalusia, on the western sea, known as Uqiyanus [Ocean].

This kingdom was made up of three distinct parts that differ from one another; the first is to the east, close to the Greek and Armenian territories, the second is the country of France, which is in the middle, and the last is Andalusia, which is to the extreme west, at the end of the inhabited world.

The capital of this entire kingdom was the great city of Rome in the territory of Almania.[3] It was founded by Romulus the Latin and was named after him. He was the first known Roman king. Rome was built some 745 years before the birth of Christ. The Latins ruled over the "limited" kingdom for 725 years following the construction of Rome, that is, until the advent of Augustus, the first of the Caesars. This Augustus defeated the king of the Greeks and annexed his kingdom to that of Rome to form a great and a single kingdom whose length from east to west, that is, from Armenia to the end of Andalusia, is about one hundred *marḥalah* [a day's journey]. And Rome became the capital of the two kingdoms and remained as such for 335 years, until Custantin [Constantine, c. A.D. 288–337], son of Helena, embraced Christianity, rejected the Sabian religion, and built a city on the Bosphorus that bears his name and is known as Constantinople. He lived in this city, which is located at the center of the Greek territory, and it became the capital of the Roman [Byzantine] kings until the present time. Ever since, these kings [Byzantine emperors] appointed Latins as their representatives to govern

Rome. These were made to function as the kings' representatives and subordinates. They were never crowned or called kings.

The Roman kings continued to run the affairs of the state and to exercise their power over the various regions of the empire for a very long time, that is, until some of their subjects, such as the Slavs, the Burgans, and others, became powerful enough to secede. Each one formed its own independent kingdom. The last one to secede was the king of Rome and that took place in A.H. 340 [A.D. 952]. Having many subjects, he felt strong and crowned himself king. Constantine, son of Leon, then king of the Rūms, sent his armies to conquer Rome, but they were defeated. Constantine corresponded with his enemy and accepted his peace. Since that time the kingdom of the Latins became different from that of the Greeks, whose western borders were pushed close to Constantinople. Then the two empires were separated by hordes of Turks who destroyed many towns and villages, so that today no one can travel between Constantinople and Rome except by sea.

The Romans of the past were Sabians until Constantine, son of Helena, the founder of Constantinople, was converted to Christianity and called on his subjects to embrace it. They all obeyed him, became Christians, and rejected the worship of idols, the glorification of temples, and other rites of the Sabian doctrine. Christianity kept growing and getting stronger until it was adopted by most of the nations neighboring the Romans, such as the Galicians, the Slavs, the Burgans, the Russians, and all the people of Egypt, such as the Copts and others, and the majority of the sects of the Sudan [black people], such as the Ethiopians, Nubians, and others.

The Romans had, in Rome and in other cities, great scholars who were knowledgeable in the various aspects of philosophy. Many people believe that the known philosophers that we have already mentioned as Greeks were Romans. But they are actually Greeks as we stated. But, because of the proximity of these two nations and the transfer of power from one to the other, the two countries became one and the two kingdoms became one, and the two peoples were diffused so that it became difficult for many to distinguish between their philosophers and to discern the history of their scholars. But those who are familiar with the annals of history and have knowledge of its details know that both nations are famous for their careful study of philosophy and that the scholars of both nations did attain a position of respectability. However, in this area no one denies, not even the Romans, that the Greeks have the superiority and the supremacy. But Allah knows best.

In the Abbāsid period of the Islamic era, there was a group of Christians and Sabians who distinguished themselves in all the branches of science. I do not know whether they were Greeks, Romans, or belonged to one of the neighboring nations. Among the Christians, there were Bakhtishūʿ and his son Jibriyl ibn Bakhtishūʿ: they were two noble physicians. Bakhtishūʿ served and treated Abū al-ʿAbbās al-Saffāḥ and those in his court. After him, he passed his services to Abū Jaʿfar al-Munṣūr [c. A.D. 712–775]. At his death, his son took over and served the kings of Banū ʿAbbās. Bakhtishūʿ was the author of many well-known medical treatises.

There was also Yūḥanna ibn Māsawayh, who was the physician of Harūn al-Rashid and al-Maʾmūn and who continued his services until the time of al-Mutawakkil. Harūn charged him with the translation of the ancient books that were found in Ankara and other Roman cities during the Muslim conquest. Yūḥanna translated many books. In the field of medicine, he has written highly respected books, among them *Kitāb al-Burhān* [Book of Proof], *Kitāb al-Baṣirah* [Book of Vision], *Kitāb al-Kamāl* [Book of Perfection], *Kitāb al-Ḥummayat* [Book of Fevers], *Kitāb al-Faṣd* [Book of Phlebotomy], *Kitāb al-Jidham* [Book of Leprosy], *Kitāb al-Aghdhiyat* [Book of Nutrition], *Kitāb Iṣlaḥ al-Aghdhiyat* [Book of the Correction of Nutrition], *Kitāb al-Maʿīdat* [Book of the Stomach], *Kitāb al-Adwiyat al-Mūsahilat* [Book on Purgatives], *al-Kinnash* [Pandects], known by the name *Muṣaghar* [Diminutive], and others.

There was also Ḥunayn ibn Isḥaq Abū Zayd, a student of Yūḥanna ibn Māsawayh, one of the best translators of the Islamic era. He had perfect knowledge of both Arabic and Greek. He learned Arabic in Baṣra[4] under the tutelage of al-Khalil ibn Aḥmad. He is the one who brought into Baghdad the book *al-ʿAyn* [The Eye]. Abū Maʿshar had mentioned in *Kitāb al-Mudhākarāt* [Book of Memoirs] that the best translators in Islam are four: Ḥunayn ibn Isḥaq, Yaʿqūb ibn Isḥaq al-Kindi, Thābit ibn Qurrah, and ʿUmar ibn Farkhān al-Ṭabarī. It is this Ḥunayn who translated and clarified the works of Hippocrates and Galen and condensed them in the best possible way. He wrote excellent treatises and remarkable documents, which include his book *al-Mantiq* [Logic], his book *al-Aghdiyat* [Nutrition], and his book *al-Adwiyat al-Mūsahilat* [The Purgatives], as well as others. Ḥunayn died during the reign of al-Mutawakkil. He fathered two sons; one was named Isḥaq and the other Dāwūd. Isḥaq [d. A.D. 902] succeeded his father as a translator. He distinguished himself in this profession and also as a mathematician. Dāwūd became a good physician.

We should also mention Masiḥ ibn Ḥakam, author of the famous *Pandects,* and Nisṭas ibn Jariḥ al-Miṣri. During the period of al-Ikhshid, there was Muḥammad ibn Ṭanj, who was a medical genius.

From among the Sabians, there was Abū al-Ḥasan ibn Thābit ibn Qurrah al-Ḥarrani [A.D. 836–911], a philosopher with a wide knowledge of all the branches of philosophy. He wrote good treatises on logic, arithmetic, geometry, astronomy, and other subjects. He was a contemporary of Yaᶜqūb ibn Isḥaq al-Kindi and Qusṭā ibn Luka. During their period of the Islamic era, those three were the most knowledgeable philosophers. During the caliphate of al-Maʾmūn, Thābit, working in Baghdad, recorded remarkable observations of the sun and collected them in a book in which he explained his theory on the solar year and its length. He also included the results of his observations of the sun while at its zenith as well as the magnitude of its movements. Thābit had a son, Sinān ibn Thābit [d. A.D. 943], who distinguished himself in arithmetic, geometry and medicine, and a grandson, Thābit ibn Sinān ibn Thābit [d. A.D. 967], a researcher in practical medicine. He lived during the period of al-Muṭiᶜ li-Allah in the princedom of Aḥmad ibn Būyh al-Dilamī al-Aqṭaᶜ, better known as Muᶜizz al-Dawlah.[5]

Chapter 10

Science in Egypt

The sixth nation [to have cultivated science] is that of Egypt. The Egyptians were the masters of a great kingdom and their glorious history goes back many centuries into the distant past. {This can be observed in the ruins of their ancient cities, old temples, and establishments, most of which are still in existence at the present time. The people of the world are in agreement that there is nothing comparable to the ruins of Egypt in any other country of the world. The history of Egypt prior to the Flood is not known, its chronicles have disappeared, but the ruins remain, such as the pyramids,[1] the temples, the caves dug in the mountains, as well as other monuments.}[2]

After the Flood the people of Egypt became a mixture of nations; there were the Copts, the Greeks, the Romans, the ʿAmaliqah [Amalek, Giants], and others. But the majority were Copts. The Egyptians became a mixture because of the various nations that took control of their country, such as the ʿAmaliqah, the Greeks, and the Romans. The various nations became a mixture, and it became difficult for the people to trace their origins; thus, by way of introduction, they were satisfied to say that they were from the country of Egypt.

The boundaries of Egypt stretch in longitude from Barqah, which is south of the Roman Sea, to Eilat on the coast of the gulf [of ʿAqaba], which extends from the sea of Ethiopia, the Zinj, India, and China, a distance of about forty days [of traveling]; and in latitude, they stretch from the city of Aswan in Upper Egypt and the regions of upper Ṣaʿīd adjoining the territory of Nubia to the city of Rashid[3] and the neighboring districts where the Nile empties in the Mediterranean Sea, a distance of thirty days.

In the distant past, the people of Egypt were Sabians; they worshiped idols and maintained temples, but were converted to Christianity at the advent of this religion. They remained as such until the Muslims conquered Egypt. At that time some of them accepted the new religion, while the rest have remained Christians until today.

The ancient Egyptians who lived before the Flood cultivated various branches of science and searched into the most complex of problems. They believed that the world "of existence and corruption" prior to the birth of the human race was populated by species of animals having strange forms and extraordinary appearances. Then came the human race, which fought and defeated all the other species until they were annihilated or dispersed in the wilderness and deserts.[4] Among these animals were the ogres and the ghouls and others as mentioned by al-Waṣyfī[5] in his book on the history of Egypt. If what was said about the ancient Egyptians is true, they were, in this respect, as far as they could be from the discipline of wisdom and the laws of philosophy.

A group of scholars claimed that all the science that was known before the Flood originated from Hermes[6] the First, who lived in Upper Egypt. This is the same person the Hebrews call Yard ibn Mahlʾyl ibn Anūsh ibn Shyth ibn Adam [Jared ibn Mahalaleel ibn Cainan ibn Enos ibn Seth ibn Adam, Genesis 5]. This is the prophet Idris—peace be upon him. It was said that he was the first to discuss stellar objects and the movements of stars. He was also the first to build temples for the glorification of God. He was the first to investigate the field of medicine and to compose for his compatriots rhythmic poems describing terrestrial as well as celestial objects.[7] It was said that he was the first to predict the Flood and to foretell that a celestial catastrophe of water and fire would strike the earth, and he became concerned that science and other forms of knowledge would be lost; so he built the pyramids that can still be found in the Ṣaʿīd of Upper Egypt. On the walls of the pyramids he drew all forms of technical equipment and devices and described all aspects of science, intending to preserve them for future generations, because he was afraid they might be lost to the world.

After the Flood, there lived in Egypt scientists who were knowledgeable in all aspects of science and philosophy, including mathematics, the physical sciences, and theology. They excelled in the compositions of talismans, Fiteh charms, burning mirrors, chemistry, and others.

In ancient times, the royal capital and the intellectual capital of Egypt was Manf [Memphis], which is about twelve miles from al-Fisṭāṭ [Old Cairo]. When Alexander built the city of Alexandria, the people were attracted to its structures, clean air, and palatable water and it became the center of education until it was conquered by the Muslims.

Soon afterward ʿUmar ibn al-ʿĀṣ built on the Nile a city that became known as Fisṭāṭ ʿUmar. The people of Egypt, Arabs and non-Arabs, came to live in it and it became the capital of Egypt and has remained such until the present day.

Among the ancient Egyptian scholars, we have Hermes the Second. He was a traveling philosopher; he journeyed through countries and visited cities; he knew their locations and the character of their inhabitants. He authored a great book on the art of chemistry and another about poisonous animals. After him came the mathematician Brūklūs [Proclus of Alexandria, c. A.D. 410–485],[8] who authored the four articles on the nature and peculiarities of arithmetic. Among their geometers and astronomers and those knowledgeable in the movements of stars, we name Theon of Alexandria [fl. A.D. 350],[9] author of a book on astronomy and *Kitāb al-Qānūn* [The Book on Laẇs]. In the book on astronomy, he included the description and the number of stars and the magnitude of the movements of the planets without any proof; this is similar to what was stated by Ptolemy in *Almagest.* In *Kitāb al-Qānūn,* he summarized the properties of planets and described their paths following the methods of Ptolemy. Then he added to it the calculation of the rotation of the skies as depicted in the talismans.

We would also like to mention Rūshum,[10] the author of great books on practical chemistry, and the scientists of Alexandria who summarized the books of Julianus [Galen] the sage and rewrote them in the form of questions and answers. The good quality of their work indicates their complete knowledge of the language and of practical medicine. They were headed by Enqylawūs, who collected from the work of Galen thirteen articles about the secrets of movements; they were written about those who engage in sexual intercourse while suffering from an incurable disease. He mentioned the outcome of such a practice and how its ills may be prevented.

Among their astronomers, we have Wālys [Valens], the author of the book known as *The Roman Yarindj,* written on the al-Mawālyd [the Newly Born] rise of the stars and the laws governing star motions. Andwarᶜar [Andrews] stated in one of his books that Valens' ten books on al-Mawālyd contain everything known about this subject and quoted Valens as saying: "I do not believe there is a science in existence that is not fully described in these books."

I do not know the definite date of any of the individual scientists of Alexandria that I have already mentioned or any additional specific information about them. Very little of their knowledge has reached us. But they have left behind a tremendous number of ruins not only in Upper Egypt but also scattered over the entire country[11] and many miraculous and splendid vases. All of this provides clear indication of the vastness of their knowledge and the discipline of their character.

Chapter 11

The Arabs: General Information

The seventh nation [to have cultivated science] is that of the Arabs, which can be divided into two groups: one group is extinct, the second still exists.

The first group comprised great nations, among them ʿĀd, Thamūd, Ṭasm, Jadīs, ʿAmaliqah, and others. Time has destroyed them and destiny annihilated them after they had exercised considerable power and enjoyed immense authority. This is a fact that no one knowledgeable in ancient history denies. Because they disappeared a long time ago, the truth about their history and traditions has been lost.

The people in the group still in existence are descendants of two grandfathers, Qaḥṭān and ʿAdnān, and share two distinct states [periods]: al-Jāhiliyah,[1] and al-Islām.

The state of the Arabs during al-Jāhiliyah is well known to many nations as a period of glory and power. Their sovereign power resided first in Qaḥṭān and then in the seven tribes that branched out from it. These tribes are Ḥimyar, Hamdān, Kindah, Lakhm, Daws, Jafnah, and Madhhaj. The great royal power belonged to Banū Ṣuwār ibn ʿAbd Shams ibn Waʾil ibn al-Ghawth ibn Ḥirān ibn Quṭn ibn ʿUrayb ibn Zuhayr ibn Ayman ibn al-Hamysiʿ ibn Ḥimyar and the other kings that followed them. Among Banū Ṣuwār there were the Jababirah and the Tabābiʿah, descendants of ancient nobility. They were people of glory and power; their well-founded kingdom conquered territories and subdued kings.

They left behind great ruins and a history full of glory well known in the East and in the West, in the North and in the South. Of their great kings, we name Yaʿrib ibn Qaḥṭān, Sabā ibn Yashjib, al-Ḥāryth al-Rāysh, Ibrahat Dhu al-Manār, ʿUmar Dhu al-Adhʿār, Abū Muʿin Ifrikus, who named Ifrikiya [Africa],[2] and Shumar Yarʿish, founder of Samarkand; there were also Tubbaʿ al-Akbar [the Great], Tubbaʿ al-Awsaṭ [the Middle], and Tubbaʿ al-Aqrun, whose name was Asʿad but became known as Abū Karb. He is the one mentioned by Abū Tammām Ḥabib ibn Aws al-Ṭāʾy[3] in a poem describing ʿAmūriyah:

> The appearance of a face whose beauty conquered
> Kisrā and cruelly ignored Abū Karb.

Finally there was Tubbaʿ al-Asghar [the Lesser], whose real name was ʿUmar Ḥassan ibn Abū Karb.

These kings had their own astrological doctrines and relied heavily on their knowledge of the behavior of the stars. Abū Muḥammad al-Ḥasan ibn Aḥmad ibn Yaʿqūb al-Hamdānī, in his book *al-ʾIklil* [The Crown], written on the history of Ḥimyar and genealogy, claimed that the kings of Ḥimyar never employed among their generals or their advisors anyone except those whose births were known to them and whose horoscopes and stars were in agreement with theirs. He further claimed that, prior to mounting an expedition, they carefully chose the favorable times that coincided with their zodiacal signs and were in agreement with the consolidation of their empire. They waited long times for these proper periods, but this provided them with victories over their enemies, adorned their conquests with glories, and made the world aware of their fame.

The kings of Ḥimyar were not interested in any stellar observations or in the study of the motions of the planets;[4] nor did they encourage the study of philosophy. In this respect they were similar to the other Arab kings who reigned during al-Jāhiliyah. As far as we know, none of them wrote anything on these subjects.

During al-Jāhiliyah, the Arabs, except for royalty, were ṭabaqatayn [two classes]: Ahl Madar and Ahl Wabar. Ahl Madar [house-dwellers] were those who inhabited villages and towns and earned their living by farming, planting palm trees and grapevines, and raising cattle. They were also known to travel long distances on business. There was no known scholar or philosopher among them.

On the other hand, Ahl Wabar [tent-dwellers] inhabited the deserts and the wilderness. They lived on the milk and the meat of camels. For grazing their cattle, when it became necessary to change pastures, they looked in the direction of lightning, the rise of clouds, and the crack of thunder. Toward these places they traveled, searching for growth where the rain had fallen. They pitched their tents in such localities as long as there was water and grass. They were always on the move, as mentioned by the poet al-Muthaqaf al-ʿAbdī, quoting his she-camel:

> She says when I put the straps on her,
> "Is this his fate as well as mine,
> Is all of life camping and moving,
> Does not he spare me or have pity on me?"

This was their condition during the heat of summer as well as spring. But during the cold winter, when the earth is serene, they would move into the rural areas of Iraq and the confines of al-Sham, where they settled close to towns and villages. There they spent the winter, suffering its hardships and stoically surviving its miseries.

During that time, they shared their food and exchanged news. They fought against oppression, helped their neighbors, and defended their women.

Their religious beliefs were diverse: Ḥimyar worshiped the sun {as indicated in what Allah mentioned in his beloved book as the hoopoe said to Solomon—peace be upon him—describing the religion of Balqys al-Ḥimyariyah: "I found out that she and her people worship the sun and not Allah." Abū Muḥammad al-Hamdānī said, "When King Solomon, son of Dāwūd [David], defeated the kings of Yemen and others, then Ḥimyar refused to worship the sun and accepted Judaism"};[5] Kinanah worshiped the moon; Tamim worshiped Aldebaran; Lakhm and Judhām worshiped al-Suha [Jupiter]; Ṭay worshiped Suhayl [Canopus]; Qays worshiped al-ʿUbūr [Sirius]; and ʾAsad worshiped ʿUṭārid [Mercury]. On the other hand, Thaqyf and ʾIyad worshiped a statue on a palm tree and they called it al-Lāt; later on, ʾIyad and Bakr ibn Waʾil worshiped the Kaʿbah of Shaddad. Ḥanifah worshiped an idol made of *ḥiys;*[6] but when they were hit by a period of famine, they ate it. In this respect, a poet said:

Ḥanifah ate their god,
during the year of famine
They did not fear of their god
his wrath or his prosecution.

Ibn Qatybah said that Christianity was practiced in Rabyʿah, Ghassān, and parts of Quḍāʿah, and Judaism in Ḥimyar and Banū al-Ḥāryth from Kaʿb and Kindah; Magianism was in Tamim and among its followers were Zararat ibn ʿAdas and his son al-Ḥājib, al-ʾAqraʿ ibn Ḥābis, and Abū Sūd, grandfather of Wakyʿ ibn Ḥassan ibn Abū Sūd. Manichaeism was in Quraysh; they adopted it from the people of al-Ḥirah. The worship of idols prevailed among the Arabs until the advent of Islam.

All the worshipers of idols, among the Arabs, believed in the unity of Allah, the Highest; but they worshiped idols as a form of the Sabian religion. They venerated the planets and the statues that represent them in temples, not as believed by those who are ignorant of the religions of nations. These are of the opinion that idol worshipers believe that the idols themselves are the gods who created the world. Neither we nor any reasonable men adhere to this view, as Allah—the Good and the Highest—said about them: "We worship them [the idols] only to bring us closer to Allah."[7]

The text of the Qurʾan on resurrection, judgment, and the prophecy of Muḥammad—may Allah's prayers and His peace be upon him—contradicts their beliefs. Their majority rejected these ideas and did not believe in future life or in final reward. They further believed that the world will never be destroyed and will not perish, even though it has been created and has a beginning. Others believed in the afterlife but held to the belief that anyone whose she-camel is slain above his grave will be resurrected riding, while one not served in this fashion will be resurrected walking. In this respect, Judhaymah ibn al-ʾAshym al-Faqʿasy, appealing to his son, said:

> O! Saʿd, when I perish, I am requesting,
>   and requests are made to the closest [relatives].
> Do not let your father walk behind them,
>   tired, falling on his hands and being confounded.
> Let him ride on a good camel,
>   do not make a mistake, do the right thing.
> I hope I will be left with a mount,
>   an animal that I can ride, when ordered to ride.

These were the religious beliefs of the Arabs.

But the knowledge that the Arabs had and were proud of was the knowledge of their language and their ability to articulate it, to compose poetry and public addresses. In addition to that, the Arabs were well informed about the history of peoples and the annals of great men. Abū Muḥammad al-Ḥasan ibn Aḥmad ibn Yaʿqūb al-Hamdānī [d. A.D. 945] said, "I have never received information about the Arabs or the non-Arabs except from the Arabs; among them the ʿAmaliqah and others, the family of al-Sumaydaʿ ibn Hūna and Khuzāʿah who lived in Makkah and who were knowledgeable in the history of the autochthonous Arabs, and the despot pharaohs as well as the history of the people of the book [Jews and Christians]." As they enter a country for trade, they learn all about its news. Also the people of al-Ḥirah, being neighbors to non-Arabs, from the time of Asaʿd Abū Karb, knew about their neighbors and their history and about Ḥimyar and their travels through the countries.[8] From them emanated most of what was reported by ʿUbayd ibn Shurayh and Muḥammad ibn al-Sāʾib al-Kalbī and al-Haytham ibn ʿAdy.[9] Similarly, Salim and the elders of Ghassān, who lived in al-Sham, had complete knowledge of the history of the Romans, Banū Israel, and the Greeks. The Arabs who lived in Baḥrayn, such as Tanūkh and ʾIyad, provided the history of Ṭasm, Bār, and Jadīs. The descendants of Naṣr ibn al-Azd, who lived in Amman and around it, provided most of the history of the kings of al-Sind and al-Hind [India] and part of the history of Persia. The people living in Ṭay supplied information about the families of Udhynah and Jarāmiqah. Finally, the inhabitants of Yemen knew the most about the history of all the nations because they lived in the kingdom of Ḥimyar, which was governed by the "traveling kings." Before mounting an expedition, their king always acquired full knowledge of the country he was about to invade and of its people.

The Arabs have excellent memory and are good narrators; they can express themselves with ease, because they were born under the zodiac that the sun traces in its passage, which contains the paths of the seven planets, thus providing information about everything.

In addition, the Arabs knew the time of rise and the time of setting of the stars and all the signs that preceded rain; they acquired this knowledge from great experience and lengthy practice; this was not due to their love for knowledge or scientific research, but it was one of the ways that helped them make a living. Abū Ḥanifah Aḥmad ibn Dāwūd al-Dynūrī, the lexicographer, wrote an excellent book about storms and winds; it contains all the knowledge that the Arabs acquired about this subject as well as related areas.[10]

This was the knowledge of the Arabs. Allah, the Highest, did not provide them with anything about the science of philosophy, nor did he prepare their character to cultivate it, and I do not know of any true Arab who became known as a philosopher except Abū Yūsuf Yaᶜqūb ibn Isḥaq al-Kindi and Abū Muḥammad al-Ḥasan ibn Aḥmad al-Hamdāni. Information about them will be presented in the proper place, Allah willing.

The country of the Arabs is known as Jazirat al-ᶜArab. It was called Jazirat [Island] because the sea surrounds it on three sides, the west, the south, and the east. To the west, there are the Gulf of Jaddah, Jār, Eilat, and the gulf that extends from the great sea, known as the Sea of Zinj and India; to the south, there is the Sea of ᶜAden, which is the great Sea of India; to the east there are the Gulf of Oman, Baḥrayn, Baṣra, and the land of Persia, which is detached from the great Indian Sea. But north of the Arabian Peninsula, one finds the southern part of al-Sham, from al-Ḥujr, which is the country of Tamūd, to Dumah al-Jandal [located between present-day Syria and Iraq] and the territories adjacent to it, which overlooks al-Samawah.

The Arabian Peninsula includes four large regions: Ḥijāz, Najd, Tahāmah, and Yemen; along its length, it extends from ᶜAden to the confines of al-Sham, a distance of about forty days' journey, and along its width, from Eilat and the Red Sea to ᶜAdhyb[11] and the neighboring plains of Iraq, a distance of about twenty-five days' journey.

Yemen was the home of the Qaḥṭāns and their place of glory from the time of Yaᶜrib ibn Qaḥṭān until the destruction of Maʾrib and its surroundings during the reign of Shumar Yarᶜish [also known as Tubbaᶜ the Great], one of the kings of Ḥimyar. This corresponds to the time of David—peace be upon him—one of the kings of Banū Israel, and to the time of Kykhsru the Third, of the third dynasty of the kings of Persia; this was some 2,060 solar years after the Flood. The reason for destruction of the dam of Maʾrib, if our sources are correct, was the small flood that devastated the dam and the city of Maʾrib as well as most of the country. The Maʾrib was then inhabited by the Azd and their allies, but upon its destruction they were scattered in the rest of the country; al-Aws and al-Khuzruj, who later became al-Anṣār,[12] went into Yathrib in the land of Ḥijāz. This is Madinat al-Nabī [the city of the Prophet]—peace be upon him. Khuzāʿah went into Makkah [Mecca] in the land of Tahāmah. Wādiʿah, Yaḥmad, Khuzām, Jadyl, Mālik, al-Ḥāryth, and al-ʿAtyk went into Amman and became known as the Azd of Amman. Māsiḥah, Mydaʿān, Lahab, ʿĀmir, Yashkur, Bāriq, ʿAli ibn ʿUthmān, Shamrān, al-Ḥujr ibn al-Hind, and Daws went into al-Surāt, which is a chain of mountains that cuts through the length of the peninsula from Yemen to the borders of al-Sham. Mālik ibn ʿUthmān ibn Daws went into Iraq. Jafnah and Banū Muḥarriq ibn ʿAmrū ibn ʿĀmir and Quḍāʿah went into al-Sham. This is not the proper place to report the migration of other Arabs from Arabia into al-Sham and into the land of Rabyʿah; we have presented what we know about this in our book *Compilations of the History of Nations from Arabs and Non-Arabs.*

This was the state of the Arabs during the period of al-Jāhiliyah; these were their beliefs, their sciences, and the way they lived. Now we shall address, Allah willing, in a most concise manner, the state of the Arabs in al-Islām.

The advent of the Prophet—may Allah's prayers and His peace be on him—took place at a time when the Arabs were subdivided and their power diminished. But Allah reunited them, brought them back together, and grouped those who were living in the Arabian Peninsula from Qaḥṭān and ʿAdnān around him [the Prophet]. They believed in him and they accepted his authority; they also rejected what they had venerated of idols and planets and singled out Allah for veneration, glorification, fatherhood, and unity. They accepted the Islamic faith and its teachings about the creation of the world and its destruction and about resurrection and reward; they agreed to act in obedience to its tenets such as prayer, fasting, tithe, *ḥajj* [pilgrimage], *jihād* [holy war], mandating the good, and forbidding the bad as well as other commandments of Islamic view.

The Messenger of Allah [the Prophet] lived only for a few years,[13] then died and was succeeded by al-Ṣaḥābah [his companions or followers]: Abū Bakr al-Ṣiddyq [the truthful], then ʿUmar al-Fārūq [the leader], then ʿUthmān al-Shahyd [the martyr], then ʿAli—Allah's prayers be upon them.[14] These caliphs pacified the country, defeated kings, and controlled kingdoms. During the reign of ʿUthmān, the kingdom of the Islam controlled vast territories and attained the high glory that the Prophet—peace be upon him—foresaw when he said, "The vastness of the earth was manifested to me; I saw its easts and its wests and the empire of my nation shall extend over everything that was manifested to me." Allah—the Highest—used the power of Islam to annihilate the Persian Empire in Iraq, Khorasan, and the rest of the Persian territories, and to annihilate the Roman Empire in al-Sham and that of the Copts in Egypt and neighboring countries. Allah—the Highest—provided the Prophet—may Allah's prayer and His peace be upon him—with the authority to govern the Arabs, first in ʿAdnān, then in Quraysh; He appointed him judge and his judgment is final and this is the way it shall be in the nations and through the ages, in accordance with what He said—power and glory are His—"Those were the times that we cause to follow one another for mankind . . ."[15]

Early during the Islamic period, the Arabs did not cultivate any of the sciences except their language and Islamic law. The only exception was the practice of medicine, which was performed by a few known Arabs whose services were needed by almost everybody. Such practice was encouraged by the Prophet—may Allah's prayer and His peace be upon him—as he said, "O creatures of Allah! attend to your health, as Allah—power and glory are His—did not create a disease without creating a remedy for it; the only exception being old age."

Among the Arab physicians, contemporary to the Prophet—peace be upon him—was al-Ḥāryth ibn Kaldah al-Thaqafy. He taught medicine in Persia[16] and in Yemen. He also played *al-ʿūd* [the lute] and lived until the time of Muʿawiyah ibn Abū Sufyān. There was also ibn Abū Rimtah al-Tamimī. He is the one who said, "I saw the seal of prophecy between the shoulders of the Prophet—peace be upon him. I said, 'I am a physician, let me remedy it,' and he answered, 'You are a companion, the physician is Allah—the Highest.'"

There was also ibn Abḥar al-Kinānī, a very able physician, who was a contemporary of ʿUmar ibn ʿAbd al-ʿAziz. ʿUmar used to send after him whenever he felt ill. There was also Khalid ibn Muʿawiyah ibn Abū Sufyān, who was skilled in both medicine and chemistry. He wrote remarkable themes and poems about the science of chemistry, demonstrating his deep understanding of this subject. This was the state of the Arabs during the rule of the Umayyads.

But when Allah—power and glory are His—wished that the power be transferred to al-Hāshimiyah [Hāshimites], there was a surge in spirit and an awakening in intelligence. The first of this dynasty to cultivate science was the second caliph, Abū Jaʿfar al-Munṣūr ʿAbd Allah ibn Muḥammad ibn ʿAli ibn al-ʿAbbas ibn ʿAbd al-Muṭṭalib ibn Hāshim.[17] He was—may Allah have mercy on him—in addition to his profound knowledge of logic and law very interested in philosophy and observational astronomy; he was fond of both and of the people who worked in these fields.

Later on when ʿAbd Allah al-Maʾmūn ibn Harūn al-Rashid ibn Muḥammad al-Mahdī ibn Abū Jaʿfar al-Munṣūr became the seventh caliph, he tried to complete what his grandfather, al-Munṣūr, had started. He searched for knowledge and extracted it from its proper sources. Because of the strength of his character and the nobility of his soul, he was able to befriend the Roman [Byzantine] emperors, shower them with precious gifts, and ask them to provide him with the books of philosophy that were in their possession; they provided him with copies of the books of Plato, Aristotle, Hippocrates, Galen, Euclid, and Ptolemy as well as those of other philosophers. He hired the ablest translators and charged them to do their best in translating these books, which they did. Then he encouraged his people to read and to study them. As a result of his efforts, a scientific movement was firmly established during his reign. And competitions among the learned prospered because of the generosity, warm treatment, and special consideration he had for scientists. He provided a special audience for scientists, listened to what they had to say, enjoyed their discussions, showered them with favors, and granted them highly respected positions. This was also the way he treated other scholars such as jurists, theologians, historians, lexicographers, poets, and genealogists.

Several learned men and scholars of al-Maʾmūn's epoch achieved high understanding of the elements of philosophy and put forth guides for research and prepared the way to education for those who came after them, so much so that al-ʿAbbāsiyah [the Abbāsid Empire] began to rival the Roman Empire when the latter was at its highest points of splendor and reunification. Then a period of decline began some three hundred years after al-Hijrah [about A.D. 913]; the authority of the caliphs was weakened and they were overcome by women and by Turks.[18] People neglected the sciences and became occupied with prevailing troubles. That continued to be the case until science was almost "elevated" [disappeared] during this period.[19] Thanks be to Allah in any case.

With the conclusion of this introduction on the history of the Arabs, we shall begin our discussion of the known scientists who lived during the reign of the Abbāsids and were Muslims, whether they were Arabs or non-Arabs.

Chapter 12
Science in the Arab Orient

The fields of science first cultivated during the Abbāsid dynasty were logic and astronomy. The first scholar to become known for his study of logic, in this dynasty, was ᶜAbd Allah ibn al-Muqaffaᶜ, the Persian orator and secretary of Abū Jaᶜfar al-Munṣūr. He translated Aristotle's three books on logic, which are the precise foundations of that science. They are the books of categories, of interpretations, and of analytics. Ibn al-Muqaffaᶜ mentioned that up to his time only the first of these three books had been translated. He also translated the introduction to the book of logic known as *Isaghuji* [Eisagoge] written by Porphyry and Mūrqūs of Tyre and others. His translation was in a simple and accessible style. He also translated the Indian book *Kalīlah wa Dimnah;* he was the first to translate it from the Persian language into the Arabic language. In addition he authored several good works; known among them is his opuscule on *adab* [literature or propriety] and politics, and another known as *al-Yatimah* [The Orphan], {later it became known as *al-Durrah al-Yatimah* [The Pearl Unique]} on submission to authority.[1]

The first person to work in astronomy during this dynasty was Muḥammad ibn Ibrahim al-Fazārī.[2] It was mentioned by al-Ḥusayn ibn Muḥammad ibn Ḥamid, better known as ibn al-Ādamī, in his *zij* [grand table] titled *Naẓm al-ᶜIqd* [Organization of the Necklace],[3] that a person originally from India came to Caliph al-Munṣūr in A.H. 156 [A.D. 773] and presented him with the arithmetic known as Sindhind for calculating the motion of stars. It contains *taᶜādyl* [equations] that give the positions of stars with an accuracy of one-fourth of a degree. It also contains examples of celestial activities such as the eclipses and the rise of the zodiac and other information. All this was in a book containing twelve chapters. Ibn al-Ādamī also stated that it was a summary of a work attributed to one of the Indian kings, named Qabghar [or Figar]. In this work calculations of star positions were carried out to an accuracy of one minute.

Al-Munṣūr ordered that the book be translated into Arabic so that it could be used by Arab astronomers as the foundation for understanding celestial motions. Muḥammad ibn Ibrahim al-Fazārī accepted the charge and extracted from it the book that astronomers called *al-Sindhind.* In the Indian language, this word means the "infinite time." This book was used by astronomers until the time of Caliph al-Maʾmūn, when it was abbreviated for him by Abū Jaʿfar Muḥammad ibn Mūsā al-Khuwarizmi,[4] who extracted from it his famous tables, which were commonly used in the Islamic world.

In addition, al-Khuwarizmi made some changes in the Sindhind system and deviated from its relations and declinations; he adopted the Persian system in formulating his equations and relied on the method of Ptolemy for determining the declination of the sun. He invented ingenious methods of approximations, but these were not enough to make up for the obvious errors in his work, which demonstrate his weakness in both geometry and astronomy. But his work was well received and was highly praised by the supporters of the Sindhind. This book is in use at the present time by those working with this type of equations of motion.

When ʿAbd Allah al-Maʾmūn ibn Harūn al-Rashid ibn Muḥammad al-Mahdī ibn Abū Jaʿfar al-Munṣūr became caliph, his noble soul craved the understanding of wisdom and the apprehension of philosophy. When the scientists of his time knew of *Almagest* and understood the construction of the observational instruments described in it, he took action. Guided by his nobility and his love for knowledge, he assembled all the scientists of his kingdom and charged them with the construction of such equipment and with its use in the study of the planets and their motions as was previously done by Ptolemy and those who came before him. Al-Maʾmūn's orders were carried out and observations began in the city of Shamāsiyah in the region of Damascus, al-Sham, in the year A.H. 214 [A.D. 829].[5] They determined the length of the solar year, the magnitude of the sun's declination, the eccentricity of its orbit, and the position of its apogee. They further studied the behavior of stars and planets until their work was interrupted by the death of Caliph al-Maʾmūn in A.H. 218 [A.D. 833]. They recorded all their observations in a book and named it *al-Raṣḍ al-Maʾmūnī* [The Observations of al-Maʾmūn].

Those who worked on this project were Yaḥyā ibn Abū Munṣūr,[6] the chief astronomer of his time, Khalid ibn ʿAbd al-Malik al-Marwarūdhī, Sanad ibn ʿAli, and al-ʿAbbās ibn Saʿīd al-Jawharī. Each one of them compiled astronomical tables that bear his name and are still in use at the present time. The observations of these scientists were the first ever performed in the Islamic Empire.

From this period until the present time, there has always been a select group of scientists, Muslims and others, who were attached to Banū ʿAbbās as well as other Muslim rulers and who worked in astronomy, geometry, medicine, and other ancient sciences. They have written important books in these fields and produced fascinating results.

Among those who became famous for their competent knowledge

in the various fields of science and philosophy, we have Yaᶜqūb ibn Isḥaq al-Kindi [c. A.D. 796–873], the philosopher of the Arabs and the son of one of their kings. He is Abū Yūsuf Yaᶜqūb ibn al-Ṣabbaḥ ibn Ismāᶜīl ibn Muḥammad ibn al-Ashʾath ibn Qays ibn Maᶜdi Karb ibn Muᶜawiyah ibn Jablah ibn ᶜAdy ibn Rabyᶜah ibn Muᶜawiyah al-Akbar [the Elder] ibn al-Ḥāryth al-Ashghar [the Younger] ibn Muᶜawiyah ibn al-Ḥāryth al-Akbar ibn Muᶜawiyah ibn Thawr ibn Murquᶜ ibn Kindah ibn ᶜAfyr ibn ᶜAdy ibn al-Ḥāryth ibn Murrah ibn Adrin Zayd ibn Yashjib ibn ᶜArīb ibn Zayd ibn Kahlan ibn Saba ibn Yashjib ibn Yaᶜrib ibn Qaḥṭān.[7] His father, Isḥaq ibn al-Ṣabbāḥ, was the governor of Kūfah during the reign of al-Mahdī and al-Rashid, and al-Ashᶜas ibn Qays was one of the companions of the Prophet—may Allah's blessings and His peace be upon him. Prior to that he was the king of Kindah, and his father Qays ibn Maᶜdi Karb was also a king over all of Kindah. He was a powerful ruler and in his praise al-Aᶜsha ibn Qays ibn Thaᶜlabah[8] wrote his four lengthy poems,[9] the first of which begins "By your life! What is the length of this time? . . ." and the second begins "Yesterday, Sumayyah chased her camels . . ." and the third begins "Did you desert Layla's family in the morning? . . ." and the fourth begins "Would you leave a beauty or stay with her? . . ."

His father, Maᶜdi Karb ibn Muᶜawiyah, was the king of Banū al-Ḥāryth al-Asghar ibn Muᶜawiyah ibn Ḥaḍramūt, while his grandfather, Muᶜawiyah ibn Jablah, was also a king in Ḥaḍramūt over Banū al-Ḥāryth al-Asghar. Muᶜawiyah ibn al-Ḥāryth al-Akbar and his father al-Ḥāryth al-Akbar and his grandfather Muᶜawiyah Abū Thawr were all kings over Maᶜd[10] in al-Mushaqqar [a fortress in Baḥrayn], in Yamamah, and in Baḥrayn. During the Islamic era, there was no man who contributed enough to the science of philosophy to merit the title "philosopher" except Yaᶜqūb al-Kindi. He wrote famous works in most scientific fields; he wrote more than fifty lengthy books and short treatises. His most famous work is a book of the unity of God, known by the title *Fam al-Dhahab* [Mouth of Gold], in which he followed Plato's doctrine and proclaimed that the universe was created at another time. He supported his opinion with faulty arguments, some of which were sophistical and some rhetorical. He also authored a book in which he refuted the philosophy of al-Mananiyah [the Manichaean], which is a heretic sect that believes in the fundamental existence of two basic principles [religious dualism], the "Good" and the "Bad."[11] He also wrote a discussion on *Fi Ma Baᶜd al-Ṭabyᶜah* [Metaphysics]. He also wrote a book titled *Fi Ithbāt al-Nubuwah* [In Support of the Prophecy] and another on the study of music known by the title *al-Mūʾnis* [The Entertainer]. He also wrote the letter *Fi Tasliyat al-Aḥzān* [In Consolation of Sorrows] and the book *Adab al-Nafs* [Principles of Morals] as well as books on logic that were well received by the public, but have limited scientific value because they are void of the analytical method, without which there is no way to determine the right and the wrong in any inquiry. No one could profit from the synthesis presented by al-Kindi in these books without proper background and premises that were not included. But with proper premises, the synthesis thought out by Yaʿqūb could be made. The premise of any inquiry cannot be obtained except through analysis. I do not know what caused al-Kindi to omit this important art. Was he ignorant of its value? Or was he loath to expose it to the public? Whatever the case may be, this was a deficiency on his part. In addition to that he wrote a great deal on a variety of scientific topics in which he exhibited improper views and presented doctrines that are far from the truth.

Among the scientists of this period, there was Aḥmad ibn Muḥammad al-Sarkhasi [d. A.D. 899].[12] He was a student of Yaʿqūb ibn Isḥaq al-Kindi; and his work in the science of philosophy shows a great deal of creativity. He is the author of remarkable works on music, logic, and other fields. His writing style is known for its excellence and brevity.

There was also Muḥammad ibn Zakariyā al-Razi [Rhazes], the unparalleled physician of the Muslims, and one of the ablest in the science of logic, geometry, and other branches of philosophy. He began his career playing *al-ʿūd* [the lute], but he soon abandoned it to study philosophy and was able to assimilate it well and to write more than one hundred books,[13] mostly in practical medicine, and some in the physical sciences and theology. But he was not able fully to understand science or to grasp its deep goal; because of that his opinions were troubled: he adopted simplistic views and insane doctrines. He squelched scholars whose ideas he was not able to apprehend and whose ways he was not able to see. He directed for some time a hospital at Rayy and then another in Baghdad. Toward the end of his life he became blind, and died around A.H. 320 [A.D. 932].

There was also Abū Naṣr Muḥammad ibn Muḥammad ibn Naṣr al-Fārābī, truly the philosopher of the Muslims. He studied logic under the tutelage of Yūḥanna ibn Ḥilān, who died in Madinat al-Salām [City of Peace, Baghdad] during the reign of al-Muqtadir.[14] Al-Fārābī surpassed all the Muslim scholars in his knowledge of logic and in his research in the field. He explained its obscurities, uncovered its secrets, and facilitated its understanding. He collected what was known or needed about logic in books that he wrote in a correct style and a delicate form. He points out in his books what was omitted by al-Kindi and others in their work on analytical methods and the branches of sciences. He clearly explains the five constituents of logic, pointing out their usefulness and the way they should be used, and how to understand the syllogistic form in each constituent. In this respect, his books are outstanding in their contents and purpose. In addition to that al-Fārābī is the author of a noble book, *Iḥṣāʾ al-ʿUlūm* [Classification and Understanding of Sciences]; there had never been a book like it and no one has tried to imitate it. The students of any of the sciences cannot do without it or proceed without its guidance.

Al-Fārābī also wrote a book on the aim of the philosophy of Plato and Aristotle, which is a testimonial to his ability as a philosopher and to his command of philosophy. This book is of great assistance in understanding the method of reasoning and the various aspects of research. In it, he uncovers the secrets of science and its fruits, one discipline after the other. And he shows how to move gradually from one science into the other. He begins with the philosophy of Plato, presents his purpose, and states what Plato has written about philosophy; then he follows this with the philosophy of Aristotle, beginning with a very enlightening and remarkable introduction in which he introduces Aristotle's philosophy step by step. Then he begins to explain the reasons for writing his books on logic and natural sciences, one book after another, until he approaches, in the copy that has reached us, the beginning of his study in theology and how it could be understood with the help of natural sciences. I do not know of a more helpful book for the study of philosophy, because it explains the common ground of all sciences and provides specifics for each one of these sciences; only by means of this book can one understand the concepts of categories and how they form the foundations of all the other sciences.

Al-Fārābī, in addition to all that, wrote two books, one on *al-ʿilm al-ʾilāhi* [theology] and the second on *al-ʿilm al-madanī* [politics]. These two books have no equal. One of them is known by the title *al-Siyasat al-Madaniyat* [Administrative Politics] and the second by the title *al-Syrat al-Fāḍilat* [The Noble Reputation].[15] In these books, the author in a superb style shows the connection between *al-ʿilm al-ʾilāhi* and Aristotle's doctrine of the six *rūhaniyah* [spiritual] principles, and how one could extract from it *al-jawāhir al-jismāniyah* [the corporal substances] that lead to order and wisdom. He also introduced the classification of humans and the power of spirit and differentiated between revelation and philosophy. He describes the various cities, the perfect and the wicked, and demonstrates that a city needs civil authority and religious laws.

Al-Fārābī was a contemporary of Abū Bushr Matta ibn Yūnus; he was younger, but more knowledgeable [than ibn Yūnus]. The books on logic written by Matta ibn Yūnus were well received in Baghdad and other Muslim cities in the East, because they are easy to understand and they contain many explanations. Matta ibn Yūnus died in Baghdad during the reign of Caliph al-Rāḍī.

Al-Fārābī died in Damascus during the reign of Prince Saif al-Dawlah ʿAli ibn ʿAbd Allah ibn Ḥamdan al-Taghlibi in the year A.H. 339 [A.D. 950].[16]

These are our most renowned scholars, who distinguished themselves in all the fields of knowledge, but we have several people who became known for their contributions to only a few of the branches of philosophy.

Among those who were known for their contribution to the science dealing with the motions of the stars and the form of the universe, we have—in addition to those already mentioned—Aḥmad ibn ʿAbd Allah al-Baghdādī, known as Ḥabash. He lived during the reigns of al-Maʾmūn and al-Muʿtaṣim and had three tables of astronomy, the first of which was prepared in accordance with the Sindhind system, in which he generally differed from the work of al-Fazārī and al-Khuwarizmi on most operations, because he adopted the use of the movements of the zodiac from the methods presented by Theon of Alexandria. This enabled him to determine the longitudinal positions of the planets. He prepared this table early in his career, when he still believed in the methods of the Sindhind. His second table, known as *al-Mumtaḥan* [The Tested], is his most famous. He prepared it after he returned to the practice of observational astronomy and included in it the motions of the planets as they were established by the tests performed in his time. The third is the smallest of the three tables and is known by the name *al-Shāh* [The King].[17] Ḥabash is also the author of a good book on the use of the astrolabe.[18]

There was also Aḥmad ibn Kathīr al-Farghānī, one of the astronomers of al-Maʾmūn and the author of *al-Madkhal ila ʿIlm Hayʾat al-Aflāk wa Ḥarakāt al-Nujūm* [The Introduction to the Science of the Shape of the Sky and the Motions of the Stars]. The book is not large, but it is of great benefit. It contains thirty chapters and presents all the material found in *Almagest* in a clear and agreeable style.

There was also Mūsā ibn Shākir[19] and his sons Muḥammad, Aḥmad, and al-Ḥasan. All four were leaders in the science of geometry and of astronomy. They all practiced observational astronomy and the measurements of the stars. Mūsā ibn Shākir was one of the best-known of al-Maʾmūn's astronomers and his sons were best known for their knowledge of the science of geometry and mechanics. On these subjects they have written many famous books, one of which is titled *Ḥiyal Banū Mūsā* [Mechanical Ruses of Banū Mūsā], which was well received by the general public.

There was also ʿUmar ibn Farkhān [Farrukhān] al-Ṭabarī [d. c. A.D. 816], a leader among translators and a knowledgeable researcher in the science of astronomy. It was stated by Abū Maʿshar Jaʿfar ibn Muḥammad ibn ʿUmar al-Balkhi in the *Kitāb al-Mudhākarāt* [Book of Discussions],[20] written to Shādhān ibn Baḥr, that Dhu al-Riyāsatayn [the one with two presidencies] al-Faḍl ibn Sahl, the vizier of al-Maʾmūn, called ʿUmar ibn al-Farkhān from his country and put him in touch with al-Maʾmūn. ʿUmar translated for al-Maʾmūn a large number of books and produced astronomical predictions that are still in the caliph's archives. He also wrote for al-Maʾmūn a large number of books about the stars as well as other fields of philosophy.

There was also Abū Jaʿfar Muḥammad ibn Sinān al-Ḥarranī, known by the name of al-Battānī [d. c. A.D. 929].[21] He was very skillful in observational astronomy, a leader in the science of geometry, celestial arrangements, and astronomical calculations. Al-Battānī prepared a remarkable astronomical table, in which he included the results of his observations of *al-nayyrayn* [the two luminaries, the sun and the moon] and corrected their motion from the error found in the book of Ptolemy known as *Almagest.* He also mentioned the distinctive motions of the five planets and included the possible corrections needed for his astronomical calculations. Some of his observations, included in his table, were made in the year A.H. 269 [A.D. 882], which corresponds to the eighth year in the reign of Caliph al-Muʿtamid.[22] I do not know of anyone in Islam who performed as well as al-Battānī in rectifying celestial observations and examining stellar movements. In addition to all that he was interested in astrology, which led him to write on this subject. Among his work is his explanation of *Kitāb al-Maqālāt al-Arbaʿa li-Baṭlymūs* [Ptolemy's Tetrabiblios].

There was also al-Faḍl ibn Ḥātim al-Nayryzī, who excelled in the science of geometry and astronomy. He is the author of famous works, among which is his book explaining *Almagest* and his book explaining the book of Euclid and a lengthy astronomical table made in accordance with the Sindhind system.

There was also al-Ḥasan ibn al-Ṣabbāḥ,[23] who prepared an astronomical table in which he established the mean positions of the stars in accordance with the Sindhind system and presented their equations of motion according to the system of Ptolemy. He also gave the declination of the sun based on the observations of his contemporaries.

There was also Muḥammad ibn Ismāʿīl al-Tanukhī,[24] the astrologer, who visited India then left it with strange information about the science of the stars and their trepidant motion as well as other subjects.

There was also ʿAli ibn Mākhur [or Amajur], a scientist very knowledgeable in the movement of the stars. He made significant astronomical observations.

There was also Abū Maʿshar Jaʿfar ibn Muḥammad ibn ʿUmar al-Balkhi, the scientist of Islam in astrology and the author of noble works and useful treatises on this subject and on the equations of motions [of the planets]. In addition to that he was the authority on the history of Persians and other people. Among his books on astrology, we have *Kitāb al-Ṭabāʾiʿ* [Book of Traits], *Kitāb al-ʾUlūf* [Book of Thousands], *Kitāb al-Madkhal al-Kabīr* [Book of the Great Introduction (to astrology)], *Kitāb al-Qiranāt* [Book of Conjunctions], *Kitāb al-Duwal wa al-Milal* [Book of Governments and Sects], *Kitāb al-Malāḥim* [Book of Poems], *Kitāb al-Aqālīm* [Book of Districts (Climates)], *Kitāb al-Hiylaj wa al-Kadkhadāh* [Book of Sense and Prediction], *Kitāb al-Maqalāt fi al-Mawālyd* [Book of Human Genetics], *Kitāb al-Nukat* [Book of Humor], *Kitāb Taḥāwil Siny al-Mawālyd* [Book of Changes in the Age of Youth], and others. Among his books on the movements of stars, we have his great astronomical tables, which are extremely beneficial, stating many astronomical facts without any proofs, and his small astronomical tables, known as the tables of *Qirānāt* [Conjunctions] and containing the mean positions of the stars and the periods of conjunctions of Saturn and Jupiter from the time of the Flood. Abū Maʿshar was a heavy wine drinker, known for his excesses. He suffered from epileptic seizures that attacked him during full moons. He was a contemporary of Jaʿfar ibn Sinān al-Battānī.

There was also al-Ḥasan ibn al-Khaṣyb, one of the leaders in the science of astrology and of astronomy. He prepared a well-known astronomical table and a good book on al-Mawālyd [The Newly Born].

There was also Aḥmad ibn Yūsuf, who wrote a commentary on Ptolemy's book on ratio and proportionality. There was also Aḥmad ibn al-Muthanna ibn ʿAbd al-Karīm, who authored *Taʿdil Zij al-Khuwarizmi* [Formulation of the Table of al-Khuwarizmi].

There was also ʿAmrū ibn Muḥammad ibn Khalid ibn ʿAbd al-Malik al-Marwarūdhī, who prepared a brief astronomical table in accordance with the proven system of his grandfather, Khalid ibn ʿAbd al-Malik al-Marwarūdhī, Yaḥyā ibn Abū Munṣūr, Sayyid ibn ʿAli, and al-ʿAbbās ibn Saʿīd al-Jawharī, who were all mentioned previously.

There was also Muḥammad ibn al-Ḥusayn ibn Ḥamid, better known as ibn al-Ādamī, who compiled the large astronomical table that was completed after his death by his student al-Qāsim ibn Muḥammad ibn Hāshim al-Madāʾinī, better known as al-ʿAlawī, who named it *Kitāb Naẓm al-ʿIqd* [Book on the Organization of the Necklace] and published it in the year A.H. 338 [A.D. 950; see note 3]. This book contains all that was known about astronomy and the calculation of the motions of the stars according to the system of the Sindhind, including certain aspects of the trepidant motions of celestial bodies, which were never before mentioned. Prior to receiving this book, we heard that it contains description of unbelievable celestial motions that obey no law. When we received this book, we persisted in its study until we understood from it what we believe was not clear to others. I pursued some of its aspects and explained them in my book written on the rectification of the motions of the stars.

There was also Abū Muḥammad al-Hamdānī, known by the name ibn Dhi al-Damynah. He was a member of Arab nobility and his genealogy follows: he is al-Ḥasan ibn Aḥmad ibn Yaᶜqūb ibn Yūsuf ibn Dāwūd ibn Sulaymān, known as ibn Dhi al-Damynah ibn ᶜUmrū ibn al-Ḥarath, ibn Munqidh ibn al-Wālid ibn al-Azhar ibn ᶜUmaru ibn Ṭāriq ibn Adham ibn Qays ibn Rabyᶜah ibn ᶜAbd ibn ᶜIlyān ibn Murrah, and this is Arhab ibn al-Diᶜām ibn Malik ibn Muᶜawiyah ibn Ṣaᶜb ibn Dūmān ibn Fykyl ibn Haytham ibn Ḥāshid ibn Nawf ibn Hamdān ibn Mālik ibn Zayd ibn Awsalah ibn Rabyᶜah ibn al-Khiyār Mālik ibn Zayd ibn Kahlān ibn Sabaʾ ibn Yashjib ibn Yaᶜrib ibn Qaḥṭān. I have extracted this genealogy from his book titled *al-ʾIklil* [The Crown], which deals with the genealogies of Ḥimyar and the periods of their kings. This is a very useful book containing ten sections: the first section deals with origins, discusses the genealogies of the Arabs and non-Arabs, and lists the descendants of Mālik ibn Ḥimyar; the second section is devoted to the descendants of al-Humaysaᶜ ibn Ḥimyar; the third section is about the nobility of Qaḥṭān; the fourth section is about ancient history from the time of Yaᶜrib ibn Qaḥṭān until the period of Abū Karb Asᶜad al-Kāmil, known as Tubbaᶜ al-Awsaṭ; the fifth section is comprised of a statement of medieval history from the time of Abū Karb until the period of Dhi Nawās; the sixth section is about modern history that extends from the time of Abū Nawās until the advent of Islam; the seventh section contains warnings against false news and impossible tales; the eighth section is about the palaces, cities, cemeteries, and poems of Ḥimyar; the ninth section is devoted to the proverbs, governments, and wars of Ḥimyar; and the tenth section is a treatment of the knowledge of Hamdān.

The book also contains good passages about the calculations of the times of conjunctions of planets and segments about natural sciences and the influence of the stars; it mentions the early concepts of the age of the universe and its creation,[25] and the various views related to the periodic cycles of the universe, the genealogy of humans, their age spans, and other subjects.

In addition to this book, al-Hamdānī authored several good works, among them his book titled *Sarāʾir al-Ḥikmah* [Secrets of Sagacity], in which he introduces the science of astronomy and the magnitude of the movements of planets, and in which he exposes in depth the science of astrology in all its forms. He is also the author of *Kitāb al-Qiwa* [Book of Strength][26] and *Kitāb Yaᶜsūb* [A Prince?], which treats the art of archery, bows, and arrows.

I found a document that bears the handwriting of al-ʾAmir al-

Andalus al-Ḥakam al-Mustanṣir ibn ʿAbd al-Raḥmān al-Nāṣir ibn Muḥammad ibn ʿAbd Allah al-ʾAmir ibn ʿAbd al-Raḥmān al-ʾAmir ibn al-Ḥakam ibn Hishām al-ʾAmir ʿAbd al-Raḥmān al-ʾAmir al-Dākhil [he who entered into al-Andalus] ibn Muʿawiyah ibn Hishām al-ʾAmir al-Muʾminīn [prince of the believers] ibn ʿAbd al-Malik al-ʾAmir al-Muʾminīn ibn Marwān ibn al-Ḥakam al-Qarshī al-Amawī, which says that Abū Muḥammad al-Hamdānī died while in jail in Sanʿāʾ in the year 334 A.H. [A.D. 946].

There was also Abū al-Ḥusayn ʿAli ibn ʿAbd al-Raḥmān ibn Yūnus al-Miṣri [from Egypt, d. A.D. 1009]; he was a specialist in astronomy with a wide knowledge of other fields of science and of poetry. He is known for rectifying the astronomical tables of Yaḥyā ibn Abū Munṣūr. He directed the Egyptians to these tables for determining the positions of the stars.

There was also al-Ḥasan ibn al-Haytham al-Miṣri, the author of treatises on burning mirrors [concave mirrors].[27] I have been informed by Qāḍi Abū Zayd ʿAbd al-Raḥmān ibn ʿĪsā ibn Muḥammad ibn ʿAbd al-Raḥmān ibn ʿĪsā[28] that he met al-Ḥasan in Egypt in the year A.H. 430 [A.D. 1038].

These are the known scholars who worked in the field of verifiable and scientific astronomy. But in the science of astrology, which is the knowledge of the effect of the stars on the rest of the world and its decomposition, the first to become known in the kingdom of Islam was Muḥammad ibn al-Haytham al-Fazārī, who has already been mentioned. He adopted the Arabic systems in his study and was followed in this method by Muḥammad ibn al-Jahm al-Barmakī, who in addition [to astrology] worked on logic. Then we have ibn Mūsāfir al-Yamānī and Khālid al-ʾAmawī and Yahyā ibn Abū Munṣūr. All of them, to a varying degree, followed the Arabic system in their work on astrology.

Among those who worked in astrology and followed non-Arabic systems such as those of the Persians, the Greeks, as well as others, we name Yaʿqūb ibn Ṭāriq, the author of *Kitāb al-Maqālāt* [Book of Statements]. In this book he discusses the births of the caliphs and kings and neglects those whose birthdates were not known to him.

There was also Ma Shāʾ Allah al-Yahudi [or al-Hindi],[29] the author of great works.

There was also Abū Sahl ibn Nahbahkhat al-Fārisī [from Persia]. He was a contemporary of al-Rashid. There was also his son, al-Faḍl ibn Abū Sahl, and Abū ʿAli al-Khayāṭ and Abū Isḥaq ibn Sulaymān al-Hāshmī, the author of a book titled *Abū Qamāsh*[30] that deals with the changes in the years of the world. There was also ʿUmar ibn Farkhān al-Ṭabarī [included earlier as an astronomer] and Abū Maʿshar Jaʿfar ibn Muḥammad ibn ʿUmar al-Balkhi and Abū Muḥammad al-Hamdānī as well as others.

Among those who became known for their work in medicine and

other fields of natural sciences, we name Isḥaq ibn ʿImrān, known by the name Sāʿh.[31] He was originally from Baghdad, but later on he went to Africa at the invitation of Ziyādat Allah ibn al-Aghlab during his reign [A.D. 903–909]. Isḥaq was a good orator and a precise scientist. He is the one who popularized medicine and philosophy in al-Maghrib [northwest Africa]. He is the author of famous books, of which we name *Kitāb Nuzhat al-Nafs* [Stroll of the Spirit], *Kitāb al-Nabaḍ* [Book of Pulse], *Kitāb al-Malankhulya* [Book of Melancholia],[32] *Kitāb al-Fasḍ* [Book on Phlebotomy], and others. Isḥaq the physician had a dispute with Prince Ziyādat Allah ibn al-Aghlab, in which he expressed his abhorrence of the ruler's despotism and simplemindedness. Al-Aghlab ordered that the veins in Isḥaq's arms be cut so that he bled to death. Then he ordered that his body be put on a cross, and it was left there until a bird built its nest in his chest cavity.

There was also Jābir ibn Ḥiyān al-Kūfi,[33] who excelled in the natural sciences, especially in the field of practical chemistry, in which he wrote many well-known treatises. In addition to that he showed some knowledge of the science of philosophy and adopted the views of the science known as *al-bāṭin* [chapter 8, note 5]. This is the sect of al-Mutaṣawifyn,[34] whose adherents among the Muslims include al-Ḥāryth ibn Asad al-Muḥāsibī [d. A.D. 857], Sahl ibn ʿAbd Allah al-Tastarī [d. A.D. 897], and others like them. I was informed by Muḥammad ibn Saʿīd of Zaragoza, also known by the name ibn al-Mashāṭ al-Asturlabī, that he had seen a manuscript in Cairo, Egypt, written by Jābir ibn Ḥiyān that deals with the use of the astrolabe. It treats one thousand questions and is without any parallel.

There was also Dhu al-Nūn ibn Ibrahim al-Aḥmymī, who was in the same *ṭabaqat* [class] as Jābir ibn Ḥiyān in his knowledge of practical chemistry and in the science known as *al-bāṭin* as well as in most of the philosophical sciences.

There was also ʿAli ibn Rabban al-Ṭabarī [c. A.D. 861],[35] the author of *al-Kinnash* [Pandect]. This work became known by the title *Firdaws al-Ḥikmah* [Paradise of Wisdom]. Ibn Rabban was the teacher of Muḥammad ibn Zakariyā al-Razi.[36]

There was also Aḥmad ibn Ibrahim ibn Khalid al-Qayrawani, better known as ibn al-Jazzar [son of the butcher]. He was well versed in medicine and well informed about what was written about it. He wrote great works on medicine as well as on other subjects. His best-known books on medicine are *Kitāb Zād al-Musāfir* [Book of the Provision of the Traveler], his book on useful medications, known by the title *al-ʾIʿtimād* [The Decision], his book on manufactured [composed] medications, known by the title *al-Baghiyat* [The Desire], and his notes on the soul and the divergence of earlier scientists on this subject. He also had a thorough knowledge of history, which led him to write a good and concise historic account that he named *Kitāb al-Taʿrif bi Ṣaḥiḥ al-Tārikh* [On Knowing the Historic Truth]. In addition to that, he was a very religious man; he enjoyed a good reputation and lived above reproach. He was independently wealthy and avoided the company of kings.

There was also ʿAli ibn al-ʿAbbās, known as ibn al-Majūsī [d. A.D. 1010]. He is the author of a book about the complete art of medicine, titled *al-Malakī* [The Royal].[37] He wrote this book for King ʿAḍad al-Dawlah Fanakhsrū ibn Rukn al-Dawlah Abū ʿAli Ḥasan ibn Būwayh al-Daylamī.[38] This is a very complete and remarkable book containing all that was known about theoretical and practical medicine. I do not know of another book equivalent to it.

These were the most famous of the Muslim scientists who lived in Iraq, al-Sham, Egypt, and Africa.[39]

Chapter 13

Science in al-Andalus

Al-Andalus, after being subdued by Banū Umayyah, became a center where a number of scholars distinguished themselves by their study of philosophy and by their understanding of many of its branches.

In ancient times, prior to the Arab occupation, al-Andalus was void of any scientific activity and none of its inhabitants became known for any scientific contribution. A few ancient inscriptions dealing with a variety of topics were found in this country, but everyone is in agreement that they were left by the kings of Rome, because al-Andalus formed a part of their empire. It remained as such, without any scientific activity, until the advent of the Muslims' conquest, which took place in the month of Ramaḍān in A.H. 92 [July, A.D. 711]. Except for the study of Islamic law and the Arabic language, the lack of interest in science persisted until the Umayyads established their authority after a period of conflict. Then persons of vitality and intelligence began their quest for science and their search for the truth, as we shall mention, Allah willing.

The religion of the people of al-Andalus was first that of the Roman Sabians, then Christianity dominated until the Muslims' conquest on the date that we have already mentioned.

The people of al-Andalus were governed by factions of various nations who took turns, one nation after the other. One of those nations was the Romans, whose agents resided in the ancient city of Ṭāliqah, located close to Ashbiliyah [Seville]. They controlled this country for a long period until they were finally defeated by al-Qūṭ [the Visigoths], who abolished the Roman influence and established the ancient city of Ṭulayṭilah [Toledo] as their capital. They reigned over al-Andalus for about three hundred glorious years, until they were defeated by the Muslims, on the date already mentioned.

The Arab kings chose Qurṭubah [Córdoba] for their capital and it remained as such until ʿAhd al-Fitnah [the Epoch of Revolts] and the civil wars against the Umayyads and the subdivision of the power in al-Andalus. The country was then governed by several rulers, as was the case in Persia during the period of al-Ṭawāʾif.

The boundaries of al-Andalus are as follows: to the south there is the Roman Sea [Mediterranean], which stretches from a place that faces Tangier known by the name al-Zuqāq [the Strait], whose width is about twelve miles and ends at the city of Ṣūr [Tyre], one of the cities of al-Sham; to the north and to the west, there is the Great Sea, known by the name Uqiyanus [Ocean], the one we call Baḥr al-Ẓulumāt [Sea of Darkness]; to the east, there is the mountain where Haykal al-Zahraḥ [the Temple of Venus] used to be. This chain of mountains extends from the Roman Sea to the Great Sea, a distance of about three days' journey, and it is the shortest of the borders of al-Andalus. Its two longest borders are those to the south and to the north; they are about thirty days' journey each. The length of its western border is about twenty days' journey.

The old city of Toledo, which was the capital of al-Qūt, is located at the center of al-Andalus at a latitude of about thirty-nine degrees and fifty minutes and a longitude of approximately twenty-eight degrees; thus it is roughly in the middle of the fifth climatic zone. Toledo, at the present time, that is, in the year A.H. 460 [A.D. 1068], is the capital of al-ʾAmir Abū al-Ḥasan Yaḥyā ibn Ismāʿīl ʿAbd al-Raḥmān ibn Ismāʿīl ibn ʿĀmir ibn Muṭarraf ibn Mūsā ibn Dhi al-Nūn, the greatest of the kings of al-Andalus. The city farthest south in al-Andalus is al-Jazirat al-Khadrāʾ [the Green Island], located on the South Sea at a latitude of thirty-six degrees. The cities having the highest latitude are those located on its northern shores, at a latitude of about forty-three degrees. Thus most of al-Andalus is located in the fifth climatic zone; only a part of it falls in the fourth climatic zone, such as Seville, Māliqah [Malaga], Córdoba, Granada, Almuriyyah [Almería], and Murciyah [Murcia]. The mountain that we have already mentioned, containing the Temple of Venus, forms the eastern borders of the country and separates al-Andalus from France, which forms a part of the immense territory of the Great Empire of the Franks. Al-Andalus is the end of the inhabited world to the west because, as we have indicated, it borders on the great Ocean Sea, after which there are no populated countries. The distance from the city of Toledo, center of al-Andalus, to the city of Rome, the capital of the Great Empire, is about forty days' journey. This is but a little information about al-Andalus.

Now let us talk about its scholars, who are the reason for mentioning al-Andalus. Toward the middle of the third century of the Hijrah calendar, which corresponds to the time of al-ʾAmir al-Khāmis [the Fifth] of the caliphs of Banū Umayyah, Muḥammad ibn ʿAbd al-Raḥmān ibn al-Ḥakam ibn Hishām ibn ʿAbd al-Raḥmān al-Dākhil ʾila al-Andalus [he who entered into al-Andalus], ibn Hishām ibn ʿAbd al-Malik ibn Marwān ibn Abū al-ʿĀṣy ibn Umayyah, some people began their scientific search in various fields and they continued to do so until about the middle of the fourth century. During that period between the two centuries, some scholars contributed to the study of mathematics and astronomy: among them, we have

Abū ʿUbayda Muslim ibn Aḥmad ibn Abū ʿUbayda al-Laythī,[1] known by the name Ṣāḥib al-Qiblah [Friend or Guardian of al-Qiblah]. He became known by that name because of his frequent and passionate prayers while facing east.[2] He was very well acquainted with the motion of the stars and their influence. In addition to that, he was a good orator and had knowledge of Islamic laws and tradition. He traveled to the East, and while in Makkah he studied with ʿAli ibn ʿAbd al-ʿAziz [d. A.D. 899],[3] and while in Egypt he studied with al-Māznī [d. A.D. 877],[4] and with al-Rabyʿ ibn Sulaymān al-Muʾadhdhin [d. A.D. 883] and then with Yūnus ibn ʿAbd al-ʾAʿla [d. A.D. 877] and Muḥammad ibn ʿAbd Allah ibn ʿAbd al-Ḥakam [d. A.D. 881] and a few others. Of Abū ʿUbayda al-Laythī, the poet Aḥmad ibn Muḥammad ʿAbd Rabbih[5] wrote:

> Abū ʿUbayda! A request for information,
> is no more than a question of him who asks.
> You refused but to be different from us;
> the opinion of those who segregate themselves
> is never on target.[6]
> Thus, the first Qiblah is changeable,[7] but you
> refused to accept anything in its place.
> You believe that Mars or Venus controls our destiny,
> or even Mercury or Jupiter or Saturn.
> You say that all the creatures are in a sphere
> that surrounds them and dictates their fate.
> The earth is spherical and the sky surrounds it
> from above and below; it appears like a point.
> Summer in the South is winter in the North
> and this condition keeps on changing.
> In December, it is cold in both San ʿāʾ and Córdoba
> and September provides them with burning heat.
> This is a guide, not an empty statement
> of the laws and it clears the word and the deed.
> As is the case of ibn Mūsā, who persisted in his error,
> making the plain[8] difficult, you mistook it
> for a mountain.
> Inform Muʿawiyah who listens to their orations
> that I renounced everything they said or did.[9]

Ibn Mūsā is Qāsim ibn Mūsā, also known as Oqashtyn [Augustine], the author, and Muʿawiyah is a Qurayshite genealogist. Abū ʿUbayda died in A.H. 295 [A.D. 908].[10]

There was also Yaḥyā ibn Yaḥyā, known as ibn al-Saminah, a native of Córdoba. He had a good knowledge of mathematics, astronomy, and medicine; he worked in most of the fields of science and excelled in language, grammar, and prosody; he was an authority on history, on *ḥadīth,* and on dialectic, and belonged to the Muʿtazilah school. He traveled to the East and returned. Yaḥyā died in A.H. 315 [A.D. 927].

There was also Muḥammad ibn Ismāʿīl, known as al-Ḥakīm [the sage]. He had a good knowledge of mathematics and logic. He was a man of fine intelligence and kind spirit. In addition to all that, he was a grammarian and a lexicographer. He died in A.H. 331 [A.D. 943].

Toward the end of the first part of the fourth century, al-ʾAmir al-Ḥakam al-Mustanṣir bi-Allah ibn ʿAbd al-Raḥmān al-Nāṣir li-Dīn Allah[11] began his effort to support the sciences and befriend the scientists. He brought from Baghdad, from Egypt, and from other eastern countries the best of their scientific works and their most valuable publications whether new or old. He began this activity during the reign of his father and continued this endeavor during the time when he was in power. His collection became equal to what the Banū ʿAbbās were able to put together over a much longer period. This was possible only because of his great love for science, his eagerness to acquire the virtue associated with it, and his desire to imitate the sage kings.

During his reign, the people became very interested in reading the books of early authors and studying and learning their doctrines; then he died in Ṣafar of A.H. 366 [September, A.D. 977].

After his death, his son Hishām al-Muʾayyad bi-Allah was proclaimed caliph. Hishām was still a boy lacking maturity, and his *ḥājib* [doorman] took over the affairs of state in al-Andalus. This *ḥājib* was Abū ʿĀmir Muḥammad ibn ʿAbd Allah ibn Muḥammad ibn ʿAbd Allah ibn Abū ʿĀmir Muḥammad ibn al-Walīd ibn Yazīd ibn ʿAbd al-Malik ibn ʾAmir al-Muʿāfirī al-Qaḥṭānī. His first action, after usurping the power of Hishām, was to take hold of the libraries of his father, al-Ḥakam. Those libraries held the previously mentioned collections of famous books as well as others; he showed these books to his entourage of theologians and ordered them to take from them all those dealing with the ancient science of logic, astronomy, and other fields, saving only the books on medicine and mathematics. The books that dealt with language, grammar, poetry, history, medicine, tradition, *ḥadīth,* and other similar sciences that were permitted in al-Andalus were preserved. And he ordered that all the rest be destroyed. Only very few were saved; the rest were either burned or thrown in the wells of the palace and covered with dirt and rocks. Abū ʿĀmir performed this act to gain the support of the common people of al-Andalus and to discredit the doctrine of Caliph al-Ḥakam. To justify his deed, he proclaimed that these sciences were not known to their ancestors and were loathed by their past leaders. Everyone who read them was suspected of heresy and of not being in conformity with Islamic laws. All who were active in the study of philosophy reduced their activities and kept, as secret, whatever they had pertaining to these sciences.

The men of talent kept to themselves what they knew about these sciences and worked in the fields that were permitted, such as mathematics, religious laws, medicine, and other similar disciplines, until the destruction of the Umayyad dynasty in al-Andalus. Toward the beginning of the fifth century of the Hijrah calendar, the country became divided by warlords and every one of them took one of its principal cities for his capital. The kings of the great civilization of Córdoba became preoccupied with these revolts to the neglect of science and learning and were finally forced to sell whatever books and furnishings were available in the palace of Córdoba; objects were sold at trivial values and at the cheapest prices. As a result, the books were scattered all over al-Andalus. This is why one may find few segments of old scientific books that were saved when the library of al-Ḥakam was destroyed during the reign of al-Munṣūr ibn Abū ʿĀmir. Everyone of the people of al-Andalus who so desired showed what was in his possession of these ancient scientific works. From then on, interest in learning the ancient sciences kept growing little by little and the capitals of al-Ṭawāʾif [sects, religions] began more and more to acquire the appearance of scientific centers. The present state, thanks to Allah, the Highest, is better than what al-Andalus has experienced in the past; there is freedom for acquiring and cultivating the ancient sciences and all past restrictions have been removed. But the kings have lost interest in these as well as other sciences; everyone is concerned with what is happening to the frontier cities and with how the heretics overpower them year after year and with the fact that their inhabitants have become too weak to defend them. All of this has led to a reduction of interest in the sciences. And in al-Andalus the people who are interested in the study of sciences have become very few.

Among those who knew mathematics and lived between the early part of al-Ḥakam's encouragement, that is, during the reign of his father, al-Nāṣir li-Dīn Allah, and the present time, we name Abū Ghālib Ḥabbab ibn ʿIbādah al-Farāyḍī; he was famous for his knowledge of the science of numbers. He lived during the reign of ʿAbd al-Raḥmān al-Nāṣir li-Dīn Allah and wrote a treatise on *al-farāʾiḍ* [religious obligations], which is still highly regarded at the present time.

There was also Abū ʾAyyūb ibn ʿAbd al-Ghāfir ibn Muḥammad, who was very proficient in the science of numbers; he also wrote a good book on *al-farāʾiḍ*. He was instructed by Aḥmad ibn Khalid, the linguist, and his *ṭabaqat.* Among his students we have Muslamah ibn Aḥmad al-Majriṭi[12] and those like him; there was also ʿAbd Allah ibn ʿUbayd Allah, known by the name al-Sarī [the rich].

He was a very able mathematician and geometer; he also wrote a famous book about *al-sabi*ᶜ [the sevens?].[13] In addition to that, he was a very able orator, grammarian, and linguist; some people attributed to him a knowledge of practical chemistry. He was held in high esteem by al-Ḥakam al-Mustanṣir bi-Allah, who wanted him in his court, but his deep religious convictions and his distaste for material wealth prevented him from accepting the caliph's generous offers.

There was also Abū Bakr ibn Abū ᶜĪsā, whose name is Aḥmad ibn Muḥammad ibn Aḥmad ibn Muḥammad ibn ᶜUmar ibn Aḥmad ibn Muḥammad ibn ᶜAbd al-Aᶜla ibn ᶜAbd al-Ghāfir ibn ᶜAbd al-Majīd ibn ᶜAbd Allah ibn Abū ᶜAbs ibn ᶜAbd al-Raḥmān ibn al-Ḥāryth al-Anṣāri, a companion of the Prophet, may Allah's blessings and His peace be upon him. He was very prominent in mathematics, geometry, and astronomy. He taught these subjects during the reign of al-Ḥakam. I have been informed by Abū ᶜUthmān Saᶜīd ibn Muḥammad ibn al-Baghūnish al-Ṭulayṭilī [from Toledo] that he had heard his teacher Muslamah ibn Aḥmad al-Majriṭi, who was a student of Abū ᶜAbs and with whom he studied geometry, recognizing his superiority not only in geometry but in all the mathematical sciences.

There was also ᶜAbd al-Raḥmān Ismāᶜīl ibn Badr, known as al-Iqlidi [follower of Euclid]. He was a leading figure in the science of geometry as well as logic. He is the author of a famous work summarizing Aristotle's eight books on logic. The son of his sister, Abū al-ᶜAbbās Aḥmad ibn Abū Ḥātim Muḥammad ibn ᶜAbd Allah ibn Harthamah ibn Dhakwān, has informed me that his uncle left al-Andalus for the East during the reign of al-Ḥājib al-Munṣūr Muḥammad ibn Abū ᶜĀmir and died there. His father, Ismāᶜīl ibn Badr, was one of the prominent citizens of Córdoba; he was noted for his knowledge of poetry and the Arabic language. He was also the director of commerce of Córdoba during the reign of Caliph al-Ḥakam, may Allah bless him.

There was also Abū al-Qāsim Aḥmad ibn Muḥammad ibn Aḥmad al-ᶜAdwī, also known as al-Ṭunayzī. He was well versed in the science of geometry and numbers. He also wrote a good book on *muᶜamalat* [business transactions].

There was also Abū ᶜUthmān Saᶜīd ibn Fatḥūn ibn Mukram al-Surqasṭī, known as al-Ḥammar [the Muleteer]. He had precise knowledge of geometry, logic, and music; he was also well acquainted with the other fields of philosophy. He wrote good works in music and one on the introduction to philosophy, which he titled *Shajrat al-Ḥikmah* [Tree of Wisdom]. He also wrote a treatise on the nature of science and how it came into existence and on the difference between what is basic and what is accidental. During the reign of al-Munṣūr Muḥammad ibn Abū ᶜĀmir, this scholar was the subject of violent persecution whose motives are well known, which led him to leave al-Andalus after being freed from jail. He died on the island of Sicily.

There was also Abū al-Qāsim Muslamah ibn Aḥmad, known by the name al-Majriṭi.[14] He was the chief mathematician in al-Andalus during his time and better than all the astronomers who came before him. He was extremely interested in astronomical observations and very fond of studying and understanding the book of Ptolemy known as *Almagest*. He wrote a good book, titled *Thimār ʿIlm al-ʿAdad* [Fruits of the Science of Numbers], which has come to mean to us the "mathematics of business transactions." He also wrote a summary of the motion of the planets from the tables of al-Battānī;[15] he also worked on the table of Muḥammad ibn Mūsā al-Khuwarizmi and changed the dates from the Persian to the Hijrah calendar and fixed the median of the positions of the stars according to the Hijrah calendar and added to it several excellent tables, but he followed al-Khuwarizmi even when he was in error without indicating the areas where such errors were committed. I have pointed that out in my book on the *Rectification of the Movements of the Stars and the Errors Committed in Observational Astronomy.*

Abū al-Qāsim Muslamah ibn Aḥmad died before the start of ʿAhd al-Fitnah, in A.H. 398 [A.D. 1008], but he left behind an excellent group of students, better than any group formed by any other scholar of al-Andalus. Among his best-known students, we name ibn al-Samḥ, ibn al-Ṣaffār, al-Zahrawī, al-Kirmānī, and ibn Khaldūn.

Ibn al-Samḥ is Abū al-Qāsim ʾAṣbagh ibn Muḥammad ibn al-Samḥ al-Mahrī.[16] He had precise knowledge of the science of mathematics and geometry and was a leader among astronomers interested in the forms of celestial spheres and the motions of the stars. In addition to all that, he was a practicing physician. He wrote several great works, among them his book *al-Madkhal ila al-Handasah* [Introduction to Geometry], in which he explained the work of Euclid, and a book, *Thimār al-ʿAdad* [Fruits of Numbers],[17] that deals with the mathematics of business. He also wrote a book titled *Ṭabiʿat al-ʿAdad* [The Nature of Number] and a great treatise on geometry in which he exhausted the study of straight, curved, and broken lines. He also wrote two books about the instrument called the astrolabe: the first is organized into two sections and describes the construction of the astrolabe; the second is about the use of the astrolabe and the result of such use, and it is organized in 130 sections. Among his other works, we have his astronomical tables, which he constructed in accordance with the Indian system known as the Sindhind; this is a large book, made up of two volumes. The first volume contains the tables and the second is a commentary on the tables. I have been informed by his student Abū Marwān Sulaymān ibn Muḥammad ibn

ʿAbs ibn al-Nāshiʾ, the geometer, that he died in the city of Granada, the capital of al-ʾAmir Ḥabbūs ibn Maksan ibn Zyrī ibn Manād al-Ṣanhājī,[18] the night of Tuesday, thirteen days before the end of Rajab in A.H. 426 [May 29, A.D. 1035]. He was fifty-six solar years old.

Ibn al-Ṣaffār is Abū al-Qāsim Aḥmad ibn ʿAbd Allah ibn ʿUmar, who is known for his precise knowledge of arithmetic, geometry, and astronomy. He lived in Córdoba, where he taught these subjects. He authored a brief table in accordance with the Sindhind system and a short, easy to understand book on the use of the astrolabe.[19] He left Córdoba a short time after the beginning of al-Fitnah to settle in the city of Denia, the capital of al-ʾAmir Mujāhid al-ʿĀmirī,[20] which is located on the eastern coast of al-Andalus. He died in it, may Allah have mercy upon him. Al-Ṣaffār taught in Córdoba a group of students who will be mentioned later; he also had a brother named Muḥammad who was famous for his construction of the astrolabe; in al-Andalus, none was better in building this instrument.

Al-Zahrawī is Abū al-Ḥasan ʿAli ibn Sulaymān. He was well versed in the science of numbers and in geometry; he also practiced medicine and wrote an excellent book on the arithmetic of transactions using the method of proof.

Al-Kirmānī [from Kirmān, Iran] is Abū al-Ḥakam ʿUmar ibn ʿAbd al-Raḥmān ibn Aḥmad ibn ʿAli al-Karmānī of the city of Córdoba. He was a scientist with a deep understanding of mathematics and geometry. I have been informed by his student, al-Ḥusayn ibn Muḥammad al-Ḥusayn ibn Ḥayy al-Tajibī, who was also a geometer and an astronomer, that he knew of no one who is as able as al-Kirmānī in understanding geometry or in the solution of its most difficult problems and demonstrating all of their parts and forms. Al-Kirmānī left al-Andalus for the East and lived in Ḥarrān of al-Jazirat [the Island].[21] There he studied geometry and medicine before returning to al-Andalus and settling in the city of Sarqasṭa [Zaragoza]. He brought with him the letters known as *Rasāʾil Ikhwān al-Ṣafā* [The Letters of the Brothers of Purity].[22] We do not know of anyone who brought them into al-Andalus before him. He also worked as a physician and made some remarkable medical *mujarrabāt* [experimentations]. He was known for practicing cauterization, amputation, incision, and ablation as well as other forms of medical surgery, but he was not knowledgeable in the science of instrumental astronomy or practical logic. This is what I was told about him by Abū al-Faḍl ibn Ḥasday ibn Yūsuf ibn Ḥasday, the Israeli.[23] He knew him well and he knew his level as a theoretical scientist, a position that no one else attained in all of al-Andalus.

Abū al-Ḥakam died, may Allah have mercy upon him, in Zaragoza in the year A.H. 458 [A.D. 1066]. He was either ninety or a little more than ninety years old.

Ibn Khaldūn[24] is Abū Muslim ʿUmar ibn Aḥmad ibn Khaldūn al-Ḥaḍramī. He was a member of the nobility of Seville. He was well acquainted with the science of philosophy and very famous for his knowledge of geometry, astronomy, and medicine. He modeled his life after the life of the great philosophers in practicing high morals, proper conduct, and noble behavior. He died in his hometown in A.H. 449 [A.D. 1057].

Among the famous students of Abū al-Qāsim Aḥmad ibn ʿAbd Allah ibn al-Ṣaffār, we have ibn Barghūt, al-Wāsṭī, ibn Shahr, al-Qarshī al-Afṭas al-Marwānī,[25] and ibn al-ʿAṭṭār.

Ibn Barghūt is Muḥammad ibn ʿUmar ibn Muḥammad ibn ʿUmar, better known as ibn Barghūt. He was known for his profound knowledge of mathematics, specializing in the study of celestial spheres, the motion of the planets, and their observation. He had, in addition to all that, a good knowledge of Arabic grammar, the Qurʾan, Arabic laws, and language and a good overview of all the other scientific fields. He was noble and kind and enjoyed a good reputation. He died, may Allah have mercy upon him, in A.H. 444 [A.D. 1053].

Al-Wāsṭī is Abū al-ʾAṣbagh ʿĪsā ibn Aḥmad, one of the authorities in the science of mathematics, geometry, and *al-farāʾiḍ* [religious obligations]. He lives in Córdoba, where he teaches these subjects. He also has some knowledge of the shapes of celestial spheres and the motion of the stars. He is still alive at the present time.

Ibn Shahr is Abū al-Ḥasan Mukhtār ibn ʿAbd al-Raḥmān ibn Mukhtār ibn Shahr al-Raʿynī. He was well versed in the science of geometry and astronomy and was a leader in his knowledge of the Arabic language, grammar, and tradition. He was an eloquent orator and a poet of great intelligence and had deep knowledge of the annals of history. He was appointed judge of Almería towards the end of the reign of Zuhayr al-ʿĀmirī[26] in A.H. 427 [A.D. 1036]. He died in Córdoba while still a judge of Almería, in A.H. 435 [A.D. 1044].

Al-Qarshī al-Afṭas al-Marwānī is Yaḥyā ibn Hishām ibn Aḥmad ibn Muḥammad ibn ʿAbd al-Malik ibn al-ʾAṣbagh, who was knowledgeable in the science of numbers and astronomy and well versed in the Arabic language and grammar [see note 25].

Ibn al-ʿAṭṭār is Muḥammad ibn Khyrah al-ʿAṭṭār, lord of the secretary Muḥammad ibn Abū Hurayrah, who was in the service of al-Ẓāfir Ismāʿīl ibn ʿAbd al-Raḥmān ibn Dhi al-Nūn.[27] He was one of the youngest of ibn al-Ṣaffār's students. He has perfect knowledge of the science of number, geometry, and *al-farāʾiḍ*. At the present time, he is teaching these subjects in the city of Córdoba. He is equally versed in practical astronomy and the movements of the stars.

Among the famous students of ibn al-Samḥ, we have Abū Marwān Sulaymān ibn Muḥammad ʿĪsā al-Nāshiʾ, who is knowledgeable in the science of number and geometry and practices the science of medicine and astrology.

We also have Abū Jaʿfar Aḥmad, better known as ibn al-Ṣaffār, the physician. He was one of the famous students of Abū Muslim ibn Khaldūn al-Qarshī, known as al-Silāḥ, whose real name is Abū Marwān ʿAbd Allah ibn Aḥmad, one of the scientists of Seville.

Among the scientists who belong in this *ṭabaqat* [class], we name ʿAbd Allah ibn Aḥmad al-Sarqasṭi [from Zaragoza]. He was well founded in his knowledge of the sciences of number, geometry, and astronomy. He remained in his hometown teaching these subjects. I have been informed about him by his student, ʿAli ibn Najdah ibn Dāwūd, the geometer, who said that he had met nobody who knew geometry and its foundations better than his teacher. I have seen a paper that ʿAbd Allah ibn Aḥmad wrote to Abū Muslim ibn Khaldūn al-Ishbilī [from Seville], in which he mentioned the errors in the Sindhind system and the relevance of its application to the movements of the stars and their formulation, and he provided some proofs of his claim. We have refuted these statements, demonstrating his error, in our book that we wrote on the *Rectification of the Movements of the Stars and the Errors Committed in Observational Astronomy.* ʿAbd Allah ibn Aḥmad died in the city of Valencia in A.H. 448 [A.D. 1056].

There was also Abū Isḥaq Ibrahim ibn Muḥammad ibn Ibrahim al-Hawazni al-Ishbilī. He was knowledgeable in the science of proof, language, and religious sects, precise and of known ability in the various fields. He died in Egypt in A.H. 420 [A.D. 1029], before reaching the age of maturity.

Among the famous companions [students] of ibn Barghūt, we have ibn al-Layth, ibn Ḥayy, and ibn al-Jallab.

Ibn al-Layth is Muḥammad ibn Aḥmad ibn Muḥammad ibn al-Layth, who had a profound knowledge of number and geometry and who studied the movement of the planets and their observation. In addition to all that, he had a clear understanding of Arabic grammar and language. He was known for his noble character and kindness. He died while serving as the judge of Shuriyūn [Surio] of the county of Valencia, in A.H. 450 [A.D. 1058].[28]

Ibn Ḥayy is al-Ḥasan ibn Muḥammad ibn al-Ḥasan ibn Ḥayy al-Tajibī of the city of Córdoba. He was versed in geometry and astronomy and formulated with passion various astronomical equations. He produced a concise astronomical table using the Sindhind system. He left al-Andalus in A.H. 442 [A.D. 1051] and entered Egypt after suffering from severe ordeals both in al-Andalus and at sea. From Egypt he traveled to Yemen and got in contact with its amir [governor], al-Ṣulayḥī,[29] who was appointed to this position by al-

Malik al-Mustanṣir bi-Allah Maᶜd ibn ᶜAli al-Ẓāhir ibn Munṣūr al-Ḥākim ibn Nizār al-ᶜAziz ibn Maᶜd al-Muᶜizz ibn Ismāᶜīl al-Munṣūr ibn ᶜAbd al-Raḥmān al-Qāʾim ibn ᶜUbayd Allah al-Mahdī,[30] whose kingdom at the present extends over parts of Africa, all of Egypt, al-Sham, the Arabian Peninsula, Ḥijāz, Tahāmah, Najd, and Yemen. Ibn Ḥayy was well received by al-ʾAmir al-Ṣulayḥī, who showered him with favors and appointed him an envoy to Caliph al-Qāʾim bi ʾAmr al-Allah[31] in Baghdad. Ibn Ḥayy made the trip in grand style and was lavishly rewarded. We were informed that he died in Yemen after his return from Baghdad and that was in either A.H. 456 or 457 [A.D. 1064, 1065].

Ibn al-Jallab is al-Ḥasan ibn ᶜAbd al-Raḥmān ibn Muḥammad, who possessed a well-founded knowledge of geometry, the shape of celestial spheres, and the motion of the stars. He also was well versed in logic and natural philosophy. At the present time he lives in the city of Almería, the capital of al-ʾAmir Muḥammad ibn Maᶜn ibn Muḥammad ibn Ṣamādiḥ al-Tajibī.

Among the leading scholars, we have Abū al-Walīd Hishām ibn Hishām ibn Khālid al-Kinānī, known by the name ibn al-Waqshi,[32] of Toledo. He is a leader in all the fields of science with a wide knowledge in all their branches. He is endowed with sure intelligence, penetrating vision, and a love for geometry and logic. He has complete knowledge of grammar, language, poetry, oration, tradition, and theology. In addition to all that, he is an eminent poet and an eloquent speaker and has deep knowledge of genealogy and history and some knowledge of all the other sciences. I met him in Toledo in A.H. 438 [A.D. 1047] and remained in his company for a very long time, learning from him and acquiring some of his knowledge. I found him to be a sea of knowledge and of perfect character; he is a man of nobility and high moral standard. He is still living, although he is over fifty years old. He told me that he was born in A.H. 408 [A.D. 1018] and became the judge of Ṭalbirah [Talavera], one of the cities in the environs of Toledo, the capital of al-ʾAmir al-Maʾmūn Yaḥyā ibn al-Ẓāfir Ismāᶜīl ibn ᶜAbd al-Raḥmān ibn Ismāᶜīl ibn ᶜĀmir ibn Muṭarraf ibn Mūsā ibn Dhi al-Nūn.[33]

We also have Abū Jaᶜfar Aḥmad ibn Khamīs ibn ᶜĀmir ibn Damj (Domingo),[34] who is also from Toledo. He worked in the science of geometry, astronomy, and medicine. He actively engaged himself in the study of literature and poetry. He was a contemporary of [a student of?] the judge Abū al-Walīd Hishām ibn Aḥmad ibn Hishām.

There was also Abū Isḥaq ibn Ibrahim ibn Lub ibn Idris al-Tajibī,[35] known by the name of al-Quwaydis. He was originally from Qalᶜat ʾAyyub [Castle of Job, Calatyud], but he left it to live in Toledo. He distinguished himself in the science of number, geometry, and *al-farāʾiḍ* and taught these subjects for a long time. He had knowledge of the shapes of celestial spheres and the movement of the stars, and I learned from him a great deal about these subjects. He also had a deep knowledge of the Arabic language and taught it for some time in Toledo. He died, may Allah have mercy upon him, the night of Wednesday, three days before the end of Rajab in A.H. 454 [A.D. 1062]. He was forty-five years old.

These are the best-known scholars who worked in the science of mathematics in al-Andalus. There were others that I have not mentioned either because they are not as good or because I did not know of them or of their contribution, although their names are well known in al-Andalus.

During our present time, there are many young scholars who have distinguished themselves in the study of philosophy and demonstrated great energy and ability to acquire a knowledge of most of its branches. Those of them who live in Toledo or around it include Abū al-Ḥasan ᶜAli ibn Khalaf ibn Aḥmar al-Ṣaydalānī [the pharmacist], Abū Isḥaq Ibrahim ibn Yaḥyā al-Naqqāsh, known by the name Walad al-Zarqāli,[36] Abū Marwān ᶜAbd Allah ibn Khalaf al-ᵓIstijī, Abū Jaᶜfar Aḥmad ibn Yūsuf ibn Ghālib al-Tamlāki, ᶜĪsā ibn Aḥmad al-ᶜĀlim [the scholar], and Ibrahim ibn Saᶜīd al-Sahlī, the constructor of astrolabes.

Among those who live in Zaragoza, we have al-Ḥājib Abū ᶜĀmir, son of al-ᵓAmir al-Muqtadir bi-Allah Aḥmad ibn Sulaymān ibn Hūd al-Jadhāmī,[37] and Abū Jaᶜfar Aḥmad ibn Jawshān ibn ᶜAbd al-ᶜAziz ibn Jawshan.

Among those who live in Valencia, we have two outstanding geometers; they are ᶜAli ibn Khalaf ibn Aḥmar al-Ṣaydalānī [the pharmacist] and Abū Jaᶜfar Aḥmad ibn Jawshān [he could have lived in both, Zaragoza and Valencia]. Also in Valencia, Abū Zayd ᶜAbd al-Raḥmān ibn Sayyid is the most knowledgeable among its scientists in astronomy and the movements of the stars.

Abū Isḥaq Ibrahim ibn Yaḥyā al-Naqqāsh, known by the name Walad al-Zarqāli, is the best in our time when it comes to astronomical observations, the study of celestial shapes, and the calculation of the movements of the stars. He is the most knowledgeable in astronomical tables and in the invention and construction of astronomical equipment.[38]

Abū ᶜĀmir ibn al-ᵓAmir ibn Hūd, although equal to his contemporaries in the study of mathematics, surpassed them all in the study of logic, natural philosophy, and theology.

Among those who specialized in the study of logic to the neglect of the other branches of philosophy, we have Abū Muḥammad ᶜAli ibn Aḥmad ibn Saᶜīd ibn Ḥazm[39] ibn Ghālib ibn Ṣāliḥ ibn Khalaf ibn Maᶜdān ibn Sufyān ibn Yazīd al-Fārisī, *mawla* [master] of Yazīd ibn Abū Sufyān ibn Ḥarb ibn Umayyah ibn ᶜAbd Shams al-Qarshī, whose ancestors came originally from the village of Mont Lisham in the district of al-Zāwiyah of the county of Unabat of the province of Lablat [Niebla] of the western parts of al-Andalus. He and his parents settled in Córdoba, where they gained wide respect and influence: his father, Abū ʿAmrū Aḥmad ibn Saʿīd ibn Ḥazm, was one of the great viziers of al-Munṣūr Muḥammad ibn ʿAbd Allah ibn Abū ʿĀmir and also of his son al-Muẓaffar, and the chief administrator of their two governments. His son al-Faqih Abū Muḥammad was the vizier of ʿAbd al-Raḥmān al-Mustaẓhir bi-Allah ibn Hishām ibn ʿAbd al-Jabbār ibn ʿAbd al-Raḥmān al-Nāṣir li-Dīn Allah. Later, he abandoned this profession and began his study of the sciences, history, and tradition. He paid special attention to the study of logic and wrote a book that he called *Kitāb al-Taqryb li-Ḥudūd al-Manṭiq* [A Book on Approaching the Limits of Logic], in which he simplified the methods of scientific acquisitions and utilized examples of legal arguments and Islamic laws. In this book he contradicted Aristotle, the founder of this science [logic], on some basic points, but his contradictions demonstrate that he did not fully understand the object of Aristotle's work. For this reason, his book is weak and contains many obvious errors. Following his work on logic, he became deeply involved in the study of Islamic law until he knew it better than anyone who lived before him in al-Andalus. He wrote on this subject a large number of treatises of high quality and noble aim. Most of these treatises dealt with the foundations and the branches of jurisprudence and were written to conform with the doctrine that he had adopted from Dāwūd ibn ʿAli ibn Khalaf al-Aṣbahānī[40] and his followers of the Ẓāhir school, although he neglected analogy and interpretation.

I have been informed by his son al-Faḍl, nicknamed Abū Rāfiʿ, that the total number of books that he wrote on jurisprudence, traditions, religious foundations, rites, and sects and his books on history, genealogy, literature, and his replies to his critics amount to about four hundred volumes, containing some 80,000 pages. In all of Islam, we do not know of a more productive author except Abū Jaʿfar Muḥammad ibn Jarīr ibn Yazīd al-Ṭabarī,[41] who was the most prolific of them all. Abū Muḥammad ʿAbd Allah ibn Muḥammad ibn Jaʿfar al-Farghānī[42] stated in his book on history entitled *al-Ṣilah* [Connection],[43] which contains detailed information about Abū Jaʿfar al-Ṭabarī al-Kabīr, that a group of his students counted the days of his life, from the time he reached maturity until his death in A.H. 310 [A.D. 922] at age eighty-six, and divided into it the total number of pages that he had written and showed that he had written fourteen pages per day. Such a feat cannot be accomplished by a person without the generous care and gracious support of his creator. In addition to all that, Abū Muḥammad ibn Ḥazm wrote large volumes on the Arabic language and its grammar, and a good section on prosody and the art of oratory. I received a letter from him, written in his own handwriting, in which he informed me that he was born after the morning prayer and before sunrise on Wednesday, the last day of Ramaḍān in A.H. 384 [A.D. 994]. He died, may Allah have mercy upon him, toward the end of Shaʿbān in the year A.H. 456 [A.D. 1064].

We also have Abū al-Ḥasan ʿAli ibn Muḥammad ibn Sydih, al-Aʿmā [the blind]. His father was also blind. He studied the science of logic for a long time and wrote many simplified works on this subject in which he followed the method of Matta ibn Yūnus. Among all the people of al-Andalus, he was the most knowledgeable in the Arabic language, grammar, and poetry. He was a man of superb memory who knew by heart several scientific works such as *al-Gharyb al-Muṣannaf* [Strange But Correct] and *Iṣlāḥ al-Manṭiq* [Correction of Enunciation].[44] He wrote great works, among them *Kitāb al-Mūḥakkam* [The Precise] and *al-Muḥyṭ al-Aʿẓam* [The Great Encyclopedia], arranged in alphabetic order, and his book *al-Mukhaṣṣaṣ* [Specialized], arranged in sections similar to *al-Gharyb al-Muṣannaf.* He also wrote a commentary on *Iṣlāḥ al-Manṭiq* and an explanation of *Kitāb al-Ḥamāsah*[45] as well as others. He died, may Allah have mercy upon him, around A.H. 458 [A.D. 1066], when he was about sixty years old. Among the scientists of al-Andalus, these are the most famous logicians.

None of the scientists of al-Andalus paid much attention to the study of natural sciences or theology, and I do not know of anyone who cultivated these sciences except Abū ʿAbd Allah Muḥammad ibn ʿAbd Allah ibn Ḥamid, better known as ibn al-Nabbāsh al-Bajjānī, who will be mentioned with the physicians Abū ʿĀmir ibn al-ʾAmir ibn Hūd and Abū al-Faḍl ibn Ḥasday al-Israeli [see note 23].

Similarly, medical science was not well understood by the people of al-Andalus and none of its scientists progressed to lead in this field. Their interest was reduced to reading some of the books that treat branches of this subject and not its foundations, such as the books of Hippocrates and Galen. They were satisfied with such superficial knowledge to cut short their study time and make their fortunes serving the kings as physicians. There were a few exceptions; they were those who disdained these aims and who chose medicine for its own sake and studied, in proper order, most of the appropriate books.

The first to become known as a physician in al-Andalus was Aḥmad ibn ʾIyas[46] of Córdoba, a very rich and a very influential person who lived at the time of al-ʾAmir Muḥammad ibn ʿAbd al-Raḥmān al-Awsaṭ [the middle one].[47] Before him, the people were medically treated by a group of Christians who were not qualified either in medicine or in any of the other sciences. They followed in their treatment one of the Christian books that was in their possession, which they called *al-Ibryshim* or *al-Ihryshim*. The word means the universal and the united.

Also during the period of al-ʾAmir Muḥammad ibn ʿAbd al-Raḥmān al-Awsaṭ, there came into al-Andalus a man from Ḥarrān who became known in al-Andalus by the name al-Ḥarranī.[48] I was never told his real name. He performed some good medical experimentation and attained great fame and a good reputation in Córdoba.

Those who came after these two physicians or were contemporary to them, but did not become as famous, include Yaḥyā ibn Isḥaq, one of the viziers of ʿAbd al-Raḥmān al-Nāṣir li-Dīn Allah during the beginning of his reign. His father Isḥaq was an able Christian physician who became famous for his medical experimentation during the reign of al-ʾAmir ʿAbd Allah [A.D. 888–912]. Yaḥyā was very intelligent and very knowledgeable about medical procedures. He was a Muslim and was highly rewarded by ʿAbd al-Raḥmān al-Nāṣir, who appointed him governor of prestigious states. He composed a medical pandect containing five volumes, in which he followed the methods used by the Romans [Christians].

There was also Saʿīd ibn ʿAbd al-Raḥmān ibn Muḥammad ibn ʿAbd Rabbih ibn Ḥabib ibn Muḥammad ibn Sālim, a confidant of al-ʾAmir Hishām al-Raḍī ibn ʿAbd al-Raḥmān al-Dākhil. He was the nephew of Aḥmad ibn Muḥammad ibn ʿAbd Rabbih [d. A.D. 940], the poet and the author of *Kitāb al-ʿIqd* [Necklace].[49] He was a noble physician and a good poet. He wrote a good medical treatise, of excellent style, in which he demonstrated his grasp of this science and of the methods used by early scientists. He was also familiar with the movements of the planets, the directions of the winds, and climatic changes. It was reported that after a surgical operation he sent after his uncle, Aḥmad ibn Muḥammad ibn ʿAbd Rabbih, the poet, asking him to come and keep him company. His uncle did not accept his invitation, so Saʿīd wrote him the following two verses:

When I have no guests or companions,
  I entertain Hippocrates and Galen.
I take their books as a remedy for my loneliness.
  They are the cure for every wound they treat.

When these two verses reached his uncle, he replied with verses, of which we choose the following:

You have accepted the company of Hippocrates and Galen,
  Because they do not eat and cost their host nothing.
At the exclusion of your relatives,
  You accepted them as friends and companions.
I believe your greed will leave you with no body,
  And after them you will accept the company of the Devil.

Saᶜīd ibn Muḥammad was a man of beautiful doctrines who avoided the company of kings. Toward the end of his life, he wrote:

> After I plunged into the study of the truth,
> And enjoyed, for a long time, the gifts of my Creator.
> And when I came close to getting into His kingdom,
> I saw those demanding wealth, but no giver.
> The age of a man is but an hour of pleasure
> That passes fast as if it were a flash of lightning.
> My soul is coming close to its departure,
> And he who drives me toward death is pressing hard.
> If I remain here or run away from death,
> To faraway places; death will catch up with me.

There were also ᶜUmar ibn Baryq, ᵓAṣbagh ibn Yaḥyā, and others. These were the physicians of al-Andalus during the period that we have mentioned, from before the time of al-ᵓAmir Muḥammad until the time when al-Ḥakam al-Mustanṣir bi-Allah expressed interest in *al-ᶜIlm* [knowledge or science] and in those interested in it.

Of the known physicians who lived in the period between the time of al-Ḥakam al-Mustanṣir bi-Allah and the present time, we have Aḥmad ibn Ḥakam ibn Ḥafṣūn, who was a noble physician of intelligence and talent. He was very perceptive. He knew logic well and was versed in all the branches of philosophy. He had connections with al-Ḥājib Jaᶜfar al-Ṣaqlabī or al-Ṣaqlī [from Sicily] and was in charge of his entourage. Al-Ḥājib put him in touch with al-Ḥakam al-Mustanṣir bi-Allah and he remained in the service of that prince until the death of al-Ḥājib Jaᶜfar. After that time, he was dropped from the medical *diwan* [guild] and was neglected until his death.

There was also Muḥammad ibn Tamlyḥ, a scholar of dignity and respect having a deep knowledge of medicine, grammar, language, poetry, and history. He served al-Nāṣir and al-Mustanṣir bi-Allah as their physician. He was also the orator of al-Ḥakam [al-Mustanṣir bi-Allah], who charged him with the supervision of the addition to the south side of the mosque of Córdoba. The work was completed under his direction and responsibility. I have seen his name written in gold and pieces of mosaic on the wall of *al-miḥrāb* [the part of the mosque reserved for the prayer leader] of that section. This structure was completed under his direction by the order of Caliph al-Ḥakam in A.H. 358 [A.D. 969].

There was also Abū al-Walīd Muḥammad ibn Ḥusayn, known as ibn al-Kinānī.[50] He was a very knowledgeable physician, an able and kind practitioner who was loved and respected by his patients. He also served al-Nāṣir and al-Mustanṣir bi-Allah.

There was also Abū ʿAbd al-Malik al-Thaqfī, an expert in medicine, mathematics, and geometry, but he spent most of his time practicing medicine and served as the physician of al-Nāṣir and al-Mustanṣir.

There were also ʿUmar and Aḥmad, the two sons of Yūnus ibn Aḥmad al-Ḥarranī. They both traveled to the East during the reign of al-Nāṣir and stayed there for ten years. They entered into Baghdad and studied the work of Galen under the tutelage of Thābit ibn Sinān ibn Thābit ibn Qurrah al-Ṣābi [of the Sabians].[51] They also entered into the service of ibn Waṣyf,[52] where they learned how to treat the eye. Then they returned to al-Andalus during the reign of al-Mustanṣir bi-Allah and this was in A.H. 351 [A.D. 962]. He invited them into his service and chose them as his own personal physicians from among all the other physicians of his time. ʿUmar died while serving al-Mustanṣir, but his brother Aḥmad continued in the service of al-Ḥakam until the end of his [al-Ḥakam's] days. His successor, Hishām al-Muʾayyad bi-Allah,[53] put Aḥmad in charge of police and commerce. But he continued his medical practice, providing remarkable eye treatment. His marvelous cures are well documented in the city of Córdoba.

There was also Muḥammad ibn ʿAbdūn al-Jabalī. He traveled to the East in A.H. 347 [A.D. 958] and lived in Baṣra [Iraq] and in Egypt, where he worked as the director of the hospital. He became a very able physician and searched great many of the fundamental roots of medicine. He was also well versed in the study of logic, which he learned from Abū Sulaymān Muḥammad ibn Ṭāhir ibn Bahrām al-Sajistānī al-Baghdādī [from Baghdad].[54] Then he returned to al-Andalus in the year A.H. 360 [A.D. 971] and entered the service of al-Mustanṣir bi-Allah and al-Muʾayyad bi-Allah as their physician. Prior to entering the field of medicine, he was a teacher of mathematics and geometry and authored a good book on *al-taksyr* [fractions].[55] I was informed by Abū ʿUthmān Saʿīd ibn Muḥammad ibn al-Baghūnish of Toledo that, when he was studying in Córdoba, there was no physician there who could be considered the equal of Muḥammad ibn ʿAbdūn al-Jabalī in the practice of medicine or in the understanding of all its branches, especially in its obscure cases.

During the time of ibn ʿAbdūn and afterward, until the end of the ʿĀmiriyah [al-ʾAmirid dynasty], lived some scientists who experimented in medicine and practiced it as a profession, but they were all at a level considerably below that of ibn ʿAbdūn. Among them there were Sulaymān ibn Ḥassan, also known as ibn Jaljal, ʿAbd Allah ibn Isḥaq, known as ibn al-Shanāʿah al-Muslamānī al-Israeli. There were also others, of whom the youngest was Muḥammad ibn al-Ḥusayn, known as ibn al-Kinānī [see note 50]. He studied medicine under the direction of his uncle, Muḥammad ibn al-Ḥusayn, and his colleagues. He worked in the service of al-Munṣūr Muḥammad ibn Abū ʿĀmir and his son al-Muẓaffar. At the beginning of ʿAhd al-Fitnah, he moved into the city of Zaragoza and settled there. He was a distinguished and able physician. He also had knowledge of logic, astronomy, and many of the branches of philosophy. The vizier Abū al-Muṭarraf ʿAbd al-Raḥmān ibn Muḥammad ibn ʿAbd al-Kabīr ibn Wāfid al-Lakhmī informed me about him, saying that he was very intelligent, a deep thinker, and endowed with the power of original productivity and induction. He was also a man of great wealth. Ibn al-Kinānī died about A.H. 420 [A.D. 1029]. He was about eighty years old. I have read some of his books, in which he says, "I learned logic from Muḥammad ibn ʿAbdūn al-Jabalī, ʿUmar ibn Yūnus ibn Aḥmad al-Ḥarranī, Aḥmad ibn Ḥafṣūn, the philosopher, Abū ʿAbd Allah Muḥammad ibn Masʿūd al-Bajjānī, Muḥammad ibn Maymūn, known by the name Mārkūs, Abū al-Qāsim Fyd ibn Najm, Saʿīd ibn Fatḥūn of Zaragoza, known as al-Ḥammār [the muleteer], Abū al-Ḥāryth al-ʾUsquf [the priest], a student of Rabyʿ ibn Zayd, priest and philosopher, ibn Maryn al-Bajjānī, and Muslamah ibn Aḥmad al-Majriṭ."[56]

Belonging in the same *ṭabaqat* [class] with ibn al-Kinānī, there was Abū al-ʿArab Yūsuf ibn Muḥammad, one of the researchers in medical science with profound knowledge of the medical field. I have been informed by the vizier Abū al-Muṭarraf ibn Wāfid and Abū ʿUthmān Saʿīd ibn Muḥammad ibn al-Baghūnish that Abū al-ʿArab was an authority on the fundamentals of medicine, with a knowledge of all its branches, and a very skillful practicing physician. I also heard from other authorities that there was no one, after ibn ʿAbdūn, equal to Abū al-ʿArab in the practice of medicine or in his knowlege of the medical field. Toward the end of his life, he was overcome by his desire for *al-khamr* [wine]. He was hardly ever sober or free from the influence of alcohol. By so doing, he prevented many people from profiting from his knowledge and ability. He died at the age of about ninety, in A.H. 430 [A.D. 1039].

From the period of these physicians until the present time there have lived several medical scientists, of whom the most famous are Abū ʿUthmān Saʿīd ibn Muḥammad ibn al-Baghūnish, who was originally from Toledo, but moved to Córdoba in his quest for scientific knowledge. He studied mathematics and geometry under the tutelage of Muslamah ibn Aḥmad and medical science under the tutelage of Muḥammad ibn ʿAbdūn al-Jabalī, Sulaymān ibn Juljul, Muḥammad ibn al-Shanāʿah, and others; then he returned to Toledo and got in touch with its prince, al-Ẓāfir Ismāʿīl ibn ʿAbd al-Raḥmān ibn Ismāʿīl ibn ʿAbd al-Raḥmān ibn Ismāʿīl ibn ʿĀmir ibn Muṭarraf ibn Dhi al-Nūn. He progressed well in the prince's court and became one of his state administrators. I met him there during the early years of the reign of al-Maʾmūn, the man of glory, ibn Yaḥyā ibn al-Ẓāfir Ismāʿīl ibn Dhi al-Nūn; by then he had abandoned the study of sciences and adhered to the study of the Qurʾan. He kept to himself at home and away from people. I found him to be a very sage man of great reputation and high moral principles, a cleanly dressed and a pious man, who had in his possession great books on the various branches of philosophy and other fields of knowledge. I came to realize by talking with him that he had studied geometry and logic and that he had precise knowledge of both fields, but he neglected this area to give special attention to the books of Galen, of which he had a private collection that he had critically corrected, thus becoming an authority on the works of Galen. He never did practice medicine and he did not have a complete understanding of diseases. He died during the morning prayer on Tuesday, the first day of Rajab, A.H. 444 [October 27, A.D. 1053]. He told me, may Allah have mercy upon him, that he was born in A.H. 369 [A.D. 980].[57] Thus, he lived to be seventy-five years old.

Among the physicians of this period, we have the vizier Abū al-Muṭarraf ʿAbd al-Raḥmān ibn Muḥammad ibn ʿAbd al-Kabīr ibn Yaḥyā ibn Wāfid ibn Muḥammad al-Lakhmī [fl. c. A.D. 1010], one of the nobility of al-Andalus and a descendant of a good family. He studied with great care the books of Galen and Aristotle as well as those of other philosophers. He distinguished himself in his study of *al-mufradah* [single ingredient] medication and knew it better than anyone else of his generation. He wrote on this subject a great book of no equal, in which he presented the content of the book of Dioscorides and the book of Galen, which were written on this same subject. He presented the material in an excellent arrangement in a single volume of some five hundred pages. He informed me that it took him some twenty years to collect, organize, state the properties and relative strengths of all the medications, and present this in his book in the fashion he deemed appropriate. As a physician, he adopted an honest approach and simple methods; he preferred not to treat his patients with medications if they could be cured with proper nutrients or something equivalent. If medication became necessary, he did not prescribe the complex if simple cures could be effective, and if compound remedies became indispensable, he preferred the use of the least complex. There are many documented cases where he cured his patients from difficult and frightful diseases with the simplest and most common medications. At the present time, al-Lakhmī is living in Toledo, and he informed me that he was born in Dhi al-Ḥijjah in A.H. 398 [August, A.D. 1008].

There was also Abū Marwān ʿAbd al-Malik, son of the jurist Muḥammad ibn Marwān ibn Zuhr al-Ishbilī [of Seville].[58] He traveled to the Middle East and lived in al-Qayrawan and in Egypt, where he studied medicine for a long time before he returned to al-Andalus and lived in the city of Denia,[59] where his fame as a physician reached most of the provinces of al-Andalus. He had medical recommendations, one of which was the forbiddance of bathing, in the belief that it encourages the growth of fungus on the body and disturbs personal behavior. This is an opinion contradicted by ancient and modern physicians and the public knows that it is erroneous. If baths are taken gradually and in proper order, they serve as a good form of exercise and a good method to open the pores and assist in getting rid of extractions and in soothing the heavy parts of chymes.

There was also Abū Muḥammad ʿAbd Allah ibn Muḥammad, known as ibn al-Dhahabī, one of those who practiced in the medical profession and read many of the books of philosophy without getting deeply into them. He was very interested in the science of chemistry, a subject that he studied intensely. He died in the city of Valencia in Jumādā II, in the year A.D. 456 [May, A.D. 1064]. I was present at his funeral. May Allah have mercy upon him.

There was also Abū ʿAbd Allah Muḥammad ibn ʿAbd Allah ibn Ḥāmid al-Bajjānī, known as ibn al-Nabbāsh.[60] He practiced the medical profession and helped cure many patients. He had a good knowledge of natural sciences and participated to some extent in the study of theology. He performed some research in the science of morals and politics and had some understanding of logic, but his knowledge of mathematics was limited. At the present time, he lives in the Murciyah [Murcia] region.

There was also Abū Jaʿfar ibn Khamīs al-Ṭulayṭilī [of Toledo],[61] who was mentioned earlier with the mathematicians. He studied the books of Galen in their proper order and adopted their content in his practice of medicine.

Among the young physicians of our time who also cultivated the study of philosophy, we have Abū al-Ḥasan ʿAbd al-Raḥmān ibn Khalaf ibn ʿAsākir,[62] who studied well the books of Galen, mostly under the direction of ʿUthmān Saʿīd ibn Muḥammad ibn Baghūnish. He is a young scientist of great character and good methods of practicing medicine. He is also very good with his hands, especially at manufacturing minute equipment. At the present time, he is working hard at understanding the science of geometry and logic. He enjoys a powerful memory and great intelligence and this, with good work and favorable conditions, may lead him to the summit in his understanding of philosophy.

In al-Andalus, the practice of astrology has met some acceptance, both in the past and at the present; there were some well-known astrologers in every period, including our own. Of the most famous astrologers during the reign of the Banū Umayyah [Umayyads], we have Abū Bakr ibn Yaḥyā ibn Aḥmad, known as ibn al-Khayyaṭ [the tailor]. He was one of the students of Abū al-Qāsim Muslamah ibn Aḥmad al-Majriṭi, who taught him the science of number and geometry. Later, he showed interest in astrology and became a well-known astrologer. In this capacity, he served Sulaymān ibn al-Ḥakam al-Nāṣir li-Dīn Allah, al-ʾAmir al-Muʾminīn [prince of the believers], during ʿAhd al-Fitnah as well as other princes, of whom the last one was al-ʾAmir al-Maʾmūn Yaḥyā ibn Ismāʿīl ibn Dhi al-Nūn. In addition to that, he practiced medicine with great care. He was kind in his judgment, a man of noble character and good reputation. He died in Toledo in A.H. 474 [A.D. 1082]. He was about eighty years old.

Among the young astrologers of our time, we have Abū Marwān ʿAbd Allah ibn Khalaf al-ʾIstijī,[63] one of those who studied astrology well and understood the old and the new books that treat the subject. I do not know anyone in al-Andalus, past or present, who has known all the secrets and marvels of this science as well as he does. He has written an excellent treatise on *Tasyrāt wa Maṭāriḥ al-Shuʿāʿāt* [The Directions and Projections of Light Rays] and some explanations of the foundation of this science. No one wrote anything like it before him. He mailed it to me from the city of Cuenca [or Fuenca].

Those are the famous Muslim scholars knowledgeable in the ancient sciences in both the East and the West.[64] I do not pretend that I know them all; it is possible that, among those who are not known to me, there are some who are better than the ones I have mentioned. Allah—the Highest—has the distinction of being all-knowing. There is no God but Him.

Chapter 14

Science of Banū Israel

The eighth nation [to have cultivated science] is Banū Israel. They were not known for their interest in philosophy, as they were occupied in the study of law and the biographies of prophets. Their rabbis know the history of prophets and of human creation better than anyone else. Their ideas were adopted by Muslim scholars such as ʿAbd Allah ibn ʿAbbās, Kaʿb al-Aḥbār, and Wahb ibn Munbih.

The Israelites have precise calculations for the history of their laws, religious obligations, and business transactions. I do not know if it is the product of their own scientists or if they have adopted it from the scientists of another race; they call this calculation *al-ʿubūr*. In it, their months are lunar and their years are either *nāqiṣah* [defective] or *mukabbasah* [compressed or leap year].[1] The *nāqiṣah* is lunar and the *mukabbasah* is solar. To correct for their historic dates, they call a period of nineteen years, from the beginning of their history, *al-maḥzūr* [the cycle]. This is the time needed to complete all fractions of years. The excess of seven months is added to specific years of *al-maḥzūr*. They are the third, the sixth, the eighth, the eleventh, the fourteenth, the seventeenth, and the nineteenth. These seven years are solar and *mukabbasah* [leap], each of which contains thirteen lunar months.[2] The rest of the years of *al-maḥzūr* are *nāqiṣah* and lunar, each of which is made up of twelve lunar months.[3] The length of their lunar year is 354 days, 8 hours, 876 minutes; one hour is made up of 1,080 minutes. The length of their solar year is 365.25 days only. Thus, their solar year is longer than their lunar year by 10 days, 21 hours, and 204 minutes.[4] The beginning of the 255th *maḥzūr* from the creation of the world corresponds, according to the Jews, to the beginning of the year 4827 from the time of Adam. This nation, from among all nations, is the cradle of prophecy and the origin of the apostolate. The majority of the prophets, may the blessings of Allah be upon them, are of Jewish origin.

They use to live in al-Sham and there they had their first and last kingdom, until the last time when they were driven out of that country by the Roman emperor Taytush [Titus]. He destroyed their kingdom and dispersed them over the entire inhabited world; there is no kingdom on earth, whether to the east or to the west, to the north or to the south, that has no Jews living in it; the only exception is Jazirat al-ᶜArab [the Arabian Peninsula], from which they were driven by ᶜUmar ibn al-Khaṭṭab, may Allah approve of him, as ordered by the Prophet, may the peace and the blessing of Allah be upon him, as he said, "There should not be two religions in the land of the Arabs." When they were scattered all over the world and were mixed with other nations, a few of them showed interest in the study of the theoretical sciences [mathematics] and in cultivating their intellectual faculties. Some of them achieved their goals in the various branches of knowledge.

Of the Jews who became famous physicians within the Muslim Empire, we have Māsirjawyh al-Ṭabib [the physician], who translated for ᶜUmar ibn ᶜAbd al-ᶜAziz[5] the medical treatise of Ahrūn al-Qiss [the cleric]. This is an excellent book and one of the best among the ancient works on this subject.

Among those who came later, we have Isḥaq ibn Sulaymān, a student of Isḥaq ibn ᶜImrān, known under the name Sāᶜh.[6] He was a distinguished physician who, in this capacity, served ᶜUbayd Allah al-Mahdī,[7] ruler of Africa. In addition, he knew logic as well as other branches of science. He lived for a long time, over one hundred years. During this long life, he never married or accumulated any wealth. He is the author of great works, among them a book on nutrition and his book on fevers that has no equal and a book on urine and *Kitāb al-Istiqṣāt* [Book of Components] and the book *al-Ḥudūd wa al-Rusūm* [Boundaries and Drawings] and his book known by the title *Bustān al-Ḥikmah* [Orchard of Wisdom], which deals with the questions of the science of theology. Isḥaq ibn Sulaymān died about A.H. 320 [A.D. 932].

Among the Jewish astrologers, we have Sahl ibn Bushr ibn Ḥabib, the author of good and well-known books in this field. He wrote *Kitāb al-Mawālyd wa Taḥāwylihā* [Births and Their Variations], *Kitāb Taḥāwyl Siny al-ᶜĀlam* [Variations of the Years of the World], and *Kitāb al-Masāʾil wa al-Ikhtibārāt* [Problems and Experimentations].

In al-Andalus, we have a group of them who practiced in the medical profession. Among them we have Ḥasday ibn Isḥaq, who served al-Ḥakam ibn ᶜAbd al-Raḥmān al-Nāṣir li-Dīn Allah. Ḥasday specialized in the art of medicine and was a leader in his knowledge of Jewish laws. He was the first to open the door and teach the Jews of al-Andalus about their culture, their laws, and their history. Before him, the Jews of al-Andalus used to resort to the Jews of Baghdad for information about their laws, their calendar, and the dates of their holy days, bringing from them the calculations for several years to help them understand the dates of their historic events and the beginnings of their years. But when Ḥasday got in touch with al-Ḥakam and was showered with his favors because of his ability, superb professionalism, talent, and manners, he asked for and received his master's assistance in acquiring whatever he wanted from the writings of the Jews of the East. Then he taught the Jews of al-Andalus the things they did not know before, thus making them abandon their difficult and costly methods.

There was also, in ʿAhd al-Fitnah, Manāḥym ibn al-Fawwāl of Zaragoza. He was a leading physician. He also knew the science of logic and many of the other branches of philosophy and wrote a very good book on the introduction to the science of philosophy and titled it *Kanz al-Muqill* [Treasure of the Poor]. He organized this book in the form of questions and answers and included in it many of the laws of logic and the foundations of natural philosophy.

Contemporary to ibn al-Fawwāl and also living in Zaragoza, we have Marwān ibn Jināḥ, who in addition to his medical practice had an immense knowledge of both the Arabic and the Hebrew languages. He is the author of a good book on *al-mufradah* [single ingredient] medication and the doses needed in treatments by weight and by volume.

There was also Isḥaq ibn Qisṭār, who was in the service of al-Mūwaffaq Mujāhid al-ʿĀmirī and his son Iqbal al-Dawlah ʿAli. Ibn Qisṭār knew the foundations of medicine and was familiar with the science of logic and the opinions of philosophers. He was a man of principles and high morals. I frequently met with him and have never seen a Jew like him, having his wisdom, truthfulness, and readiness to help. He was a leader in his knowledge of the Hebrew language and the Jewish laws. He was also a rabbi and died in Toledo in A.H. 448 [A.D. 1056]. He was seventy-five years old and had never been married.

Among the Jewish scholars, there were a few who showed some interest in certain branches of philosophy. In this group, we have Sulaymān ibn Yaḥyā, known by the name ibn Jubayr [Avicebron of the Middle Ages]. He lived in Zaragoza and was fond of the science of logic. He was a scientist of keen intelligence and sound judgment. He died while still in his early thirties, around A.H. 450 [A.D. 1058].

Among their youths who live in our era, we name Abū al-Faḍl Ḥasday ibn Yūsuf ibn Ḥasday[8] of the city of Zaragoza and of the Jewish nobility in al-Andalus; he is a descendant of the Prophet Moses, peace be upon him. Abū al-Faḍl studied the sciences in the proper order, adopting the best methods. He learned with precision the Arabic language, its rhetoric, and the composition of poetry. He excelled in the science of number, geometry, and astronomy. He understood the art of music and tried to practice it. He showed deep interest in the science of logic and practiced the various methods of research and observations in this field. Then he elevated himself to the study of the natural sciences and began by studying Aristotle's book of *al-Kiyān* [The Cosmos of Physics] until he understood it well, then he took to the study of *Kitāb al-Samāʾ wa al-ʿĀlam* [Book of the Sky and the World]. This is when I left him in A.H. 458 [A.D. 1066], while he was uncovering the unknown. If Allah provides him with His protection and he lives long, he shall perform well in the field of philosophy and all its branches. He is still young and has not attained manhood, but Allah, the highest, provides generously to whomever He pleases. He is all-powerful.

These are the Hebrews of our land who have excelled in the science of philosophy. But the Jewish scientists who specialize in Jewish laws are too numerous to count both in the East and in the West. Of those in the East, we name Saʿīd ibn Yaʿqūb al-Fayyūmī[9] [of the Fayyum district of Egypt], Abū Kathyr Yaḥyā ibn Zakariyā al-Kātib al-Ṭabarānī, Dāwūd al-Qūmshī, Ibrahim al-Tustarī, as well as other Jewish rabbis who talk about the various religious sects and practice the science of dialectic, discussion, and argument.

Of those who lived in al-Andalus, we have Abū Ibrahim Ismāʿīl ibn Yūsuf al-Kātib [the author], known by the name al-Ghazal,[10] who worked in the service of al-ʾAmir Bādīs ibn Habbush al-Ṣanhājī, the king of Granada and its provinces. He was the director of the state. He knew the Jewish laws and how to defend and protect them more than any other Jewish scholar of al-Andalus. He died in A.H. 448 [A.D. 1056].

This is what I remember of the names of the scholars of the nations and the small selections of their works and their annals. To Allah alone we offer thanks, may His benedictions and peace be upon the last of His prophets, our lord Muḥammad, and upon the members of his family and his companions.

—Completed by the grace of Allah, the Highest—

Notes

Ṣāʿid al-Andalusī

1. Al-Andalus (Andalusia or Andalucía in Spanish) is the name the Arabs bestowed on southern Spain. Al-Andalusī is the Arabic form for Andalusian—someone from al-Andalus.
2. In Spain this was ʿAhd al-Fitnah [the Epoch of Revolts], A.D. 1010–1070/A.H. 400–462, when warlords and princes divided the country into several feudal dominions and fought among themselves to extend their domains. Nevertheless, all forms of intellectual pursuit continued to flourish in spite of the dismemberment of the state and the political instability.
3. A critical biography of Ṣāʿid was recently prepared by Martin Plesser (1956).
4. On that date, the Arab forces led by Mūsā ibn Nuṣayr and his general Ṭāriq ibn Ziyād crossed what became known as the Strait of Gibraltar from Africa into Spain. The name Gibraltar is a corruption of the Arabic words Gibal Ṭāriq [Mountain of Ṭāriq].
5. Most of the Arab scholars of al-Andalus made similar trips to further their education. It was their way of paying homage to the center of learning of the time.
6. Ismāʿīl ibn ʿAbd al-Raḥmān al-Ẓāffir was born and raised in Sontebria, which was then ruled by his father. When a state of turmoil arose in Toledo, its inhabitants asked the prince of Sontebria for his help. Al-Ẓāffir was sent to Toledo; he settled its problems and became its prince.
7. This prince was also known by the name Abū Zakariyā al-Maʾmūn (Shaykhū, Blachère, and Bū-ʿAlwan). But in *Ṭabaqāt al-ʾUmam,* Ṣāʿid refers to him as al-ʾAmir Abū al-Ḥasan Yaḥyā ibn Ismāʿīl ibn ʿAbd al-Raḥmān ibn Ismāʿīl ibn ʿĀmir ibn Muṭarraf ibn Mūsā ibn Dhi al-Nūn.
8. The Arabs then used the word *mamlakah* [kingdom] to mean a country regardless of its form of government. The root of the Arabic word signifies possession.
9. Al-Fatiḥ Muḥammad ibn Yūsuf was born in Madinat al-Faraj [Guadalajara]. After a trip to the Orient, he returned to his place of birth, where he died in A.D. 1059/A.H. 451.
10. Abū Walīd al-Waqshi was born in Huecas in A.D. 1017/A.H. 407. He was a well-known poet, author, geometer, and judge. He died in A.D. 1096/A.H. 489.
11. Al-Tajibī is referred to by Ṣāʿid in *Ṭabaqāt al-ʾUmam* as a geometer,

astronomer, and mathematician. He left Spain for Egypt in A.D. 1051/ A.H. 442 and died in Yemen upon his return from Baghdad in A.D. 1064/ A.H. 456.

Introduction

1. The commonly used Arabic word *ṭabaqāt* is the plural form of *ṭabaqat,* whose meaning varies from "class" when referring to people or nations to "layer" when discussing geological formations. Various forms of the word are used to denote floors in buildings as well as trays of food.
2. Régis Blachère performed comprehensive research on the subject that enabled him to identify and sometimes correct names, dates, book titles, and so forth. But his knowledge of the Arabic language was somewhat limited. Many passages in the French translation do not correctly convey the Arabic text. Blachère based his translation on Father Shaykhū's work, which is also flawed.
3. Beside *ṭabaqāt,* the words "al-Sham" and "Sabians" are frequently encountered in the text. Al-Sham is the name the Arab bestowed on Greater (or geographical) Syria, the territory between the Mediterranean Sea and Mesopotamia. Ṣāʿid uses the word "Sabians" in a loose sense, as a mask word to cover a variety of beliefs. In general, it is used to denote a religion whose adherents worshiped and venerated idols and heavenly bodies. This religion is mentioned in the Qur'an: "Lo! Those who believe and those who are Jews, and the Sabians, and the Christians and the Majus and the Idolaters . . ." (Qur'an, al-Ḥajj [the Pilgrimage], XXII, 17).
4. In addition to the copy at the Chester Beatty Library, whose number is 3950, there is a copy in each of the following locations: National Library, Paris, No. Arabe 6735; Oriental Library, Saint Joseph University, Beirut, No. 158; ʿAshir Afandi Library, Istanbul, No. 668; Köprülü Library, Istanbul, No. 1105; and British Museum, No. Add. 1622. There may be others.
5. We are unable to detect any difference between the style or the language of the section devoted to the science of the Hebrews and the rest of the manuscript. Therefore, it is more likely that the copy at the Chester Beatty Library is incomplete than that the remainder of the work was added in a later addition.
6. Lutz Richter-Bernburg, "Ṣāʿid, the Toledan Tables," in King and Saliba (1987:374).
7. For more information about this scholar and astronomer, see *DSB,* XIV, 592–593, and IX, 40.
8. Toomer (1968:1, 1970:306).
9. Raymond of Marseilles (fl. first half of the twelfth century A.D.) borrowed his astronomical data from al-Zarqāli, preferring his figures to those of Ptolemy. In so doing, he appears to have given Latin astronomers one of their first contacts with what would be, for some four centuries, the standard method for determining the positions of the planets (*DSB,* XI, 321).

10. The natural philosophers of the Middle Ages were rarely if ever specialists; they were not interested in differentiating among the various branches of knowledge. A savant was valued not for his detailed knowledge in any specialized field, but for his complete mastery of culture in encyclopedic proportions.
11. Ṣāᶜid intentionally wrote in a concise and highly condensed style. This was a common practice during that period. But his brevity makes certain passages rather difficult to understand.
12. Sarton (1927–1953: III, 1620).
13. Chejne (1974: 176).
14. Sarton (1927–1953: I, 777).
15. King and Saliba (1987: 390).
16. Ibn Khaldūn (1958: I, 81n; II, 214n, 365n; III, 126n, 127n, 247n).
17. Lawrence (1976: 92).
18. Khan (1980: 153).
19. Mahdi (1957: 143–144).
20. Chaube (1969: 102, 112, 114, 238, 239).
21. These omissions are most probably the result of the unawareness in al-Andalus scholarly circles of the scientific development that took place in al-Mashriq after the middle of the tenth century. The inclusion in *Ṭabaqāt al-ᵓUmam* of the two scholars Abū al-Ḥusayn ᶜAli ibn ᶜAbd al-Raḥmān ibn Yūnus and al-Ḥasan ibn al-Haytham to the neglect of others may be due to the fact that they were both residents of Egypt, which is on the route from Spain to Makkah. Al-Ḥajjaj, returning to Andalusia, passed through Egypt and carried the news of its scientists. It is safe to add that it took almost half a century after the death of Ṣāᶜid for the science of the Islamic East to be incorporated into al-Andalus scholarship (King and Saliba 1987: 373 and references therein).
22. *DSB*, I, 5.
23. Sarton (1927–1953: II, pt. 1, 66).
24. Haskins (1927: 7–19).

1. The Seven Original Nations

1. The original manuscript was not divided into chapters. This was first performed by Shaykhū and then by Blachère. We have adopted it for the convenience of readers.
2. This was the traditional way Arab authors began their work, by first stating their names as if quoting themselves. Both Shaykhū (1912: 5) and Bū-ᶜAlwan (1985: 33) write "Ṣāᶜid said" instead of "Ṣāᶜid wrote."
3. Most Islamic scholars divided the world into seven nations (see, for example, Khalidī, 1975).
4. These are not present-day Syrians; the Sirianiyūns lived in the region prior to the Arab invasion.
5. Al-Sham is the name the Arabs bestowed on the territories north of the Arabian Peninsula. The city of Damascus is still referred to as al-Sham.
6. The complete title of Masᶜūdī's famous book is *Murūj al-Dhahab wa Maᶜādin al-Jawhar* [Prairies of Gold and Quarries of Jewels]. This en-

cyclopedia of medieval Islamic knowledge is usually referred to by its short title, *Murūj al-Dhahab.*

7. This is not to be confused with the Great Northwest quadrant of ancient civilization, which included North Africa, western Asia, and all of Europe (Breasted 1936:129–133).
8. "Sabians" is probably derived from the Sabians of Ḥarrān, northwestern Mesopotamia. Ṣāʿid uses the term to designate primitive religions (see introduction, note 3).
9. Ancient astronomers knew of five planets (Mercury, Venus, Mars, Jupiter, and Saturn). These planets plus the sun and the moon were referred to collectively as the seven planets. To the Babylonians, the Indians, and the Chaldeans, they became important divinities. They sang the praise of each god on a particular day. It took them seven days to worship all seven deities; hence the seven days of the week and their names (Sunday, Monday . . . and Saturday for Saturn).

   The commonly used phrase "I am in seventh heaven" could have had its origin in Babylonian mythology, and the same could probably be said of Aristotle's seven celestial spheres.

2. *The Two Categories of Nations*

1. In addition to the singular and the plural forms, Arabic words have a dual form. The *ayn* ending means two; hence *ṭabaqatayn* means two *ṭabaqāt*s.
2. It seems that Ṣāʿid is using the Arabic word *ʾumam* to mean peoples.
3. Hajūj and Majūj were the ancient names of the northwestern region of Siberia.
4. Khazars are the people in the Caucasus region.
5. Ṭilsāns are probably the inhabitants of Tilsit [Sovetsk] in Russia.
6. Kazakhs are the Turko-Tartar Muslims who came to central Asia.
7. The Alains are the Scythian people who lived in Russia and the Black Sea regions.
8. The Arabic word for black people is the same as the name of the country, Sudan.
9. Nubians are the people who lived in northwest Africa in the region of the Nile Valley.
10. Ghana or Gana formed a vast empire in West Africa between the fourth and thirteenth centuries.
11. Some of these names have not been included in the various copies of the manuscript or in Blachère's translation. In a few instances, we were unable to correlate the Arabic and English names.

3. *Nations Having No Interest in Science*

1. Instead of eastern, Bū-ʿAlwan (1985:41) writes "drinking regions." The two Arabic words *mashāriq* [eastern] and *mashārib* [drinking regions] are not readily distinguishable.
2. Both Shaykhū (1912:8) and Bū-ʿAlwan (1985:41) write "education" in place of "tacticians."
3. Boga or Bega is probably that part of Egypt located between the Nile and the Red Sea (*EI,* 1st ed., I, 705).

4. Nations That Cultivated the Sciences

1. Some of these proverbs or variations of them are commonly used in parts of the Arab world. A few of them are verses from well-known poems. We believe that in this section copyists were not faithful to the original manuscript: they have added some old sayings and changed others. Some copies contain additional proverbs, such as "more vanity than a rooster," "more courage than a lion," "more loyalty than a dog" or "a cat," "more timid than a pigeon," and so forth. Blachère experiences some difficulties in translating these proverbs into French.
2. Ṣāʿid mentions in chapter 1 that all the people of the world formed seven nations. Some of these nations were later subdivided, thus accounting for eight nations that have cultivated science. For example, the Chaldeans became Chaldeans, Arabs, and Hebrews and the Greeks became Greeks and Romans. Ṣāʿid is thus using the word *ʾumam* to mean peoples as well as nations.
3. In reality, the style of the manuscript is too concise. In many instances the ideas are so interwoven and the expressions are so terse that a reader senses that clarity has been sacrificed for the sake of brevity.

5. Science in India

1. As we have seen, the Arabs referred to all countries as kingdoms and called all heads of states kings. Thus the sultans of Turkey and the emperors of Rome were called kings.
2. Sudan, the name of the country, is derived from the Arabic word for black; Ṣāʿid may be referring to the color rather than the country.
3. This refers to the Indian systems of numbers, which they passed to the Arabs via Persia (Karpinskii 1965:47). This system is still being used worldwide.
4. Hindus believe in one God, having different manifestations. "To what is one, sages give many different names" (Ṛg-Veda I, 164, 46).
5. Brahmins are the highest of the Hindu castes. "Sabians" here denotes the lower three castes: the Kṣatriya, the Vashya, and the Sudra.
6. The validity of this statement is in doubt. It could mean they are in agreement as to the law of prophecy that prohibits the slaughter, maltreatment, or consumption of animals. Blachère also makes a similar statement (1935:45).
7. Hindus believe in a cyclic universe (Vāyu-Purāṇa I, 7, 72) that evolves and decays. Like the Greeks, they believe that the universe is made of four elements—earth, air, fire, and water. At the end of the cycle, these four elements cease to exist in their manifest form (Ṛg-Veda X, 129, 1–4).
8. This is a very important statement considering the barriers of distances and languages of that period and is possibly a deciding factor when it comes to the content of this manuscript. The exchange of scientific information across international boundaries was meager and at times dangerous.
9. The original titles of these books are mentioned by al-Birūnī (A.D. 973–1050/A.H. 362–441), the Islamic philosopher who accompanied

Maḥmūd Gajni during the invasion of India and lived there for some thirteen years, A.D. 1017–1030/A.H. 407–421 (Sachau 1964:I, 153, 312, 316; II, 7, 19, 48, 90). Shaykhū writes *al-izjir* instead of Arjabhar (1912:13).

10. Al-Fazārī is the Arab scholar who translated the Sindhind, the Ārjbahd, and the Ārkand from Farsi into Arabic.
11. All these scholars flourished in Baghdad during the Abbāsid era.
12. According to al-Birūnī (Sachau 1964:I, 153), the word *Sindhind* is derived from *Siddhānta,* which means "straight, not crooked or changing." The actual meaning of *Siddhānta* is "rule."
13. These numbers add up to 4.32 billion years. This is the length of one cycle of creation and is referred to in Hindu scriptures as the Mahayuga, the long period (see Vishnu-Purāṇa VI, 3; Manu-Saṁhita I, 61–74).
14. For more information about the Indian astronomical system, the reader is referred to *DSB* (Supplement, XV, 533) and references therein.
15. The original Sanskrit title of this book is *Panchatantra* [Five Rules]. *Kalīlah wa Dimnah* is still a rather popular book in the Arab world. For more about ibn al-Muqaffaʿ, see chapter 12.
16. Hindus spread sawdust or sand on tablets and performed their calculations, writing with their fingers. This technique became known in the Arab world by the name Ḥisāb al-Ghubār [arithmetic of dust or dustboard arithmetic], not Ḥisāb al-Ghiyar [change] as stated by Shaykhū (1912:14).
17. Kanka or Manka left India for Baghdad and entered into the service of Harūn al-Rashid. He became his astrologer and translated some of the works of Indian scholars for him.
18. For added clarification, see Pingree (1968) and chapter 12.

6. *Science in Persia*

1. Al-Ṭawāʾif is the name by which the Arabs recognized the Persian dynasty that ruled the country after the destruction of the empire of Alexander of Macedon. The word means tribes or religious factions.
2. This period was first proposed by Abū Maʿshar in his book *Zij al-Ḥaḍārāt* and is based on the Sindhind and Ārkand astronomical systems.
3. Abū Maʿshar was born near Balkh in Khorasan, which was a center of learning and a meeting place of Indian, Chinese, Arab, and Persian scholars. He moved to Baghdad during the reign of al-Maʾmūn (A.D. 813–833/A.H. 197–218) and served as an astrologer in his court under the tutelage of al-Kindi (c. A.D. 796–873/c. A.H. 179–259). For more detailed information, see chapter 12.
4. *Jāmāseb* [Djamasp] could be the name of the author rather than the title of the book. Someone by that name preached the Magian religion after the death of Zoroaster and authored a book on alchemy (ibn al-Nadīm 1871–1872:354).
5. Budasaf studied astronomy and astrology in India, returned to Persia, reintroduced astrology, and converted the Persians from the worship

of idols to the worship of heavenly bodies (Pingree 1968:4–5, 16, 19).

6. Vishtaspa was the king of Bactra [Balkh]. See, for example, Huart (1972:168–169).

7. Missing from this chapter are some of the great Persian scholars who flourished during the Abbāsid era. The two best known are ibn Sīnā and ʿUmar al-Khayyam. Ibn Sīnā [Avicenna] was born in Afshana, near Bukhara [now Uzbek, USSR] in A.D. 980/A.H. 369 and died in Hamadan [Iran] in A.D. 1037/A.H. 428. His precocity was extraordinary: at the age of ten, he knew the Qurʾan by rote; at sixteen, he was practicing medicine; and at seventeen, he was appointed physician at the Samanid court. After serving in various administrative positions, he was named vizier of Shams al-Dawlah.

Ibn Sīnā wrote some 270 works in a number of widely divergent fields. His major work in philosophy, *al-Shifāʿ* [The Cure (from ignorance)] is an immense four-part encyclopedia of logic, physics, mathematics, astronomy, and metaphysics. He also wrote a compendium of his work called *al-Najāt* [The Deliverance].

His major medical work, *al-Qānūn*, already mentioned in the introduction, is about one million words long and was translated into Latin by Gerard of Cremona (Milan, A.D. 1473/A.H. 878). It is appropriate to judge the work of ibn Sīnā by comparing it to the work of al-Razi (chap. 12); ibn Sīnā was the better philosopher and al-Razi was the better physician. Ibn Sīnā borrowed extensively from al-Razi's work *al-Ḥāwi* (*DSB*, Supplement 1, XV, 494–500).

ʿUmar al-Khayyam [the tentmaker] (c. A.D. 1020–1110/c. A.H. 410–503) was born in Nishapur, but spent most of his youth in Balkh. He was a leading algebraist, an astronomer, and a philosopher. His greatest mathematical contribution is his *Risālah fīʾil-Barāhyn ʿala Masāʾil al-Jabr wa al-Muqabalah* [Treatise on the Proofs of the Problems of Algebra and "Balancing"]. This title gives the impression that the author intended his book to be a follow-up to al-Khuwarizmi's famous book *Kitāb al-Jabr wa al-Muqabalah* (chap. 12, note 4). While working as director of the astronomical observatory in Isfahān, al-Khayyam performed most of his work in astronomy. Based on a cycle of thirty-three years, he calculated the mean length of the solar year and obtained a value of 365.2424 days, which is only 0.00021 days or about 18 seconds longer than the actual value.

Al-Khayyam wrote five philosophical treatises; because of his free thinking, he was accused of atheism. In 1859, Edward Fitzgerald published his English "translation" of seventy-five quatrains of the famous *Rubaʿiyat al-Khayyam.* A Persian version of the *Rubaʿiyat* containing over 1,000 quatrains was recently published.

7. *Science of the Chaldeans*

1. From Genesis 10: Noah begat Ham and Ham begat Cush and Cush begat Nimrod—the mighty hunter; and the beginning of his kingdom was Babel. "Therefore is the name of it called Babel; because the Lord did there confound the language of all the earth" (Genesis 11:9). The

name Babel actually comes from the Akkadian phrase *bab ili* [the gate of the god].

2. Qur'an, Surat al-Naḥil [Bees], XVI, 26.
3. The *dirāᶜ* is a unit of length, which was widely used in the Middle East before the advent of the metric system. A *dirāᶜ* is not the cubit often mentioned in the Old Testament. A cubit is the distance from the elbow to the tip of the fingers, or about 45 cm, while the *dirāᶜ* is the distance from the shoulder to the tip of the fingers, or about 70 cm. According to Ṣāᶜid's account, the tower of Babel must have been 3,500 meters high—a clear impossibility. It is now believed that the ziggurat of Babylon was less than 100 meters in height.
4. The Persians entered Babylon in 538 B.C., during the reign of Belshazzar, son of Nebuchadnezzar (Breasted 1936:263 and the Book of Daniel, chap. 5).
5. Ṣāᶜid is repeating an old Arab tale that states that there were three Hermes: the first or the old Hermes lived in Egypt and taught arts and sciences and predicted the Flood; the second, Hermes of Babylon, was a contemporary of the Greek philosopher Socrates; while the third, Hermes of Egypt, was a student of Pythagoras. Some historians believe that the last two Hermes were one person who was born in Babylon, then traveled to Egypt (ibn al-Nadīm 1871–1872:352).
6. The Greeks gave the name Hermes Trismegistus [Thrice-greatest] to the Egyptian god Thoth, a reputed author and the source of the Hermetic writing. He was the patron of all the arts dependent on writing, including medicine, astronomy, magic, and alchemy. Medieval chemistry was often called "hermetic science." There is also a Greek god named Hermes; he is the son of Zeus and Maia, the daughter of Atlas. He was associated with the protection of cattle and sheep. Ṣāᶜid could have obtained his information from Greco-Roman sources.
7. Hipparchus collected the previous astronomical works of Greeks and Babylonians and observed the periodic nature of eclipses. He realized the importance of the almanac and classified 1,080 stars according to their brightness and locations.
8. This book must have been known by two different titles: *al-Baraydaj* and *al-Zaharj* (ibn al-Nadīm 1871–1872:269; and al-Qifṭī 1903:261).
9. The Chaldeans' worship of their divinities (the planets) gave rise to their practice of astrology, which has survived to the present day. This belief led them to carry out continuous observations of the skies. In the twenty-third century B.C., in the days of kings Sumer and Akkad, they recorded an observation of the eclipse of the moon that has now been calculated by modern astronomers to have occurred in 2283 B.C. Around 500 B.C., some 2,000 years before the invention of the telescope, a Chaldean astronomer named Nabu-rimannu calculated the length of the year to be 365 days, 6 hours, 15 minutes, and 41 seconds, only 26 minutes and 55 seconds too long. Around 400 B.C., another Chaldean astronomer, named Kidinnu, proved that there is a difference between the length of the year as measured from equinox to equinox and as measured between successive arrivals of the earth at

its nearest point to the sun. Without realizing it, Kidinnu discovered the precession of the earth's axis (Breasted 1936:212–214). The Chaldean astronomers compiled records of astronomical observations for over 360 years, by far the longest uninterrupted astronomical observations on record. By comparison, the longest series of modern observations began in A.D. 1675 in Greenwich, England. This great interest in observational astronomy and the Chaldean sexagesimal system of numbers resulted in the 360-degree circle, the sixty-minute hour, and so forth.

10. The original title of Ptolemy's book was *Megale.* Harūn al-Rashid had the book translated into Arabic. The translator added the definitive article *al-* and in admiration changed the title from *Megale* [great] to *Megiste* [greatest] (al-Yaʿqūbī 1883). Eventually it appeared in Arabic under the title *Kitāb al-Majisṭi,* which in medieval times was transformed to *Magasiti.* At present, the accepted title of this famous book is *Almagest.*

11. In the eleventh century, knowledge of the early Babylonian Empire was limited; only recent excavations have brought to light the extent of this ancient civilization. Thus Ṣāʿid gave a brief account of the contributions of the second Babylonian Empire (the Chaldeans) without mentioning the Epic of Gilgamesh (c. 2800 B.C.), the Epic of Atrahasis (c. 1650 B.C.), or the law code of Hammurabi (c. 1750 B.C.).

8. *Science in Greece*

1. Dhu al-Qarnayn is the name given to Alexander the Great by Arab historians. It has nothing to do with a probable muscular conformation that gave his head a tilt toward the left shoulder; rather it refers to the title he assumed after the assassination of Darius—emperor of Europe and Asia. Thus Dhu al-Qarnayn symbolizes the two continents.

2. After Alexander's death (June, 323 B.C.) in Babylon, his empire was subdivided; the Ptolemies took over Egypt and the eastern Mediterranean countries and established a dynasty of fourteen kings who ruled the area for some three hundred years. The renaissance of Alexandria was accomplished mainly by the first two Ptolemies: Ptolemy, son of Lagos, and his son Ptolemy the second, who ruled until 246 B.C. During the reign of the Ptolemies Alexandria became a center of science and knowledge. In Alexandria the Ptolemies paid salaries to and supported the largest group of natural philosophers known in the ancient world. This is probably the first scientific institution fully supported by a government. Thus the scientists of the Hellenistic age, especially the remarkable group at Alexandria, became the founders of systematic scientific research, and their books formed the body of scientific knowledge for nearly two thousand years, until the revival of science in modern time (Sarton 1954:6–20).

3. This sentence is missing from Blachère (1935:58).

4. Lukman is a sage, who is featured in the Qur'an as a wise father giving pious admonitions to his son. He is also celebrated for his wisdom by pre-Islamic Arab poets (*EI,* V, 811).

5. *Al-bāṭiniyah* is an Arabic word that means within, inside, or hidden from view. Here it is used to denote ideas that are implied, but not clearly stated; they are in a sense mysterious and not easy to apprehend. *Al-bāṭin* and *al-ẓāhir* are two Sufi technical terms that mean "esoteric" and "exoteric," respectively. Blachère translates *bāṭin* to mean "hidden" (1935:59) and *ẓāhir* to mean "exterior" (1935:141).
6. Al-ᶜAllaf al-Baṣrī (and not al-Miṣri as in Shaykhū 1912:22) is a learned theologian of the Muᶜtazilah school. He died around A.D. 845/A.H. 230.
7. The chronology is in error. Empedocles was a Pythagorean philosopher and was born after Pythagoras. He was censured by the members of the Pythagorean school for publishing some of the Pythagoras' teachings.
8. Here Blachère adds a passage reiterating the discussion between Socrates and the king, then indicates that the passage is not in the original manuscript (1935:61).
9. This is but a sentence from a lengthy passage that Blachère adds to the manuscript (1935:62).
10. Al-Masᶜūdī was a tenth-century Islamic historian. He was born in Baghdad but lived most of his life in Egypt. He was a prolific author, having written some thirty-seven works (Khalidī 1975).
11. Here Blachère adds a very lengthy passage describing how Aristotle helped Philippus choose Alexander as his successor (1935:67–68).
12. This is Alexander the physician, whose best-known work is a book on the diseases of the eye and their cures (al-Qifṭī 1903:55).
13. Porphyry was a student of Plato. He wrote a biography of Plato and explained some of his ideas in his book *Madkhal ila al-Maᶜqulāt* [An Introduction to Reasoning]. He was fond of prognostication and astrology (ibn al-ᶜIbri 1890:78; ibn al-Nadīm 1871–1872:253).
14. For more about al-Kindi, see chapter 12.
15. As his name indicates, Luka was a Christian Syrian scholar. He traveled to Greece for his education; upon his return to Syria, he was summoned to Baghdad to translate books. He was a physician and a translator (ibn al-ᶜIbri 1890:149; al-Qifṭī 1903:262–263).
16. Blachère (1935:69) translates *al-Nātiq wa al-Ṣāmit* to mean "the Reasonable and the Unreasonable" and *Nisbat al-Akhlāṭ* to mean "Report of Temper."
17. Hippocrates is also known as the father of medicine. His writings became the nucleus of a collection of medical treatises written by a number of authors, but all attributed to him; all still bear his name. He reflected in medicine the enlightenment of his age in Greece and maintained, against religious views, that diseases must be treated as subject to natural laws. He studied medicine in the Pythagorean tradition. Present-day physicians still take the Hippocratic oath.
18. This section is missing from Shaykhū (1912:28) and from Blachère (1935:69).
19. Although born in Pergamum, Asia Minor, Galen settled in Rome and became the personal physician of Emperor Marcus Aurelius and L. Verus. He was perhaps the first to propose that human arteries contain blood and not air (Galen 1952:163).
20. Apollonius of Perga was a famous mathematician. His book *Conics*

was translated into Arabic during the reign of Caliph al-Maʾmūn.

21. What little is known about the life of Euclid, the great geometer, has come to us through Arabic translators. He is Euclid ibn Nuqatrus ibn Barniqus, known as the father of geometry. The Greek title of his book on geometry is *al-Astrushia* [Foundation of Geometry]. This book was used in Europe as a textbook for many centuries. Euclid was a scholar of Greek ancestry whose country was Syria, whose hometown was Tyre, and whose profession was carpentry (al-Qifṭī 1903:62–64). But Euclid flourished in Alexandria—hence the name Euclid of Alexandria.

22. Archimedes was a native of Syracuse, Sicily. One of his famous feats was an arrangement of a series of pulleys and levers that enabled the king, by turning a light crank, to move a large three-masted ship standing on the dock and launch it into the water. After witnessing such a feat, the people of Syracuse believed his proud boast: "Give me a place to stand and I will move the earth." He was also an accomplished inventor. He was able to prove to the king that one of his crowns was not pure gold, because he discovered the principle of determining the portion of loss of weight when an object is immersed in water. This principle, which bears his name, is still a part of most introductory physics texts.

23. Simplicius was one of the most famous advocates of Neoplatonism in the sixth century. He belonged to the school of Ammonius Hemiae in Alexandria.

24. Hipparchus was previously mentioned as a Chaldean astronomer (chapter 7). It seems that neither Ṣāʿid nor Shaykhū (1912:19, 21), Blachère (1935:55, 72), and Bū-ʿAlwan (1985:69, 88) realized they were writing about the same astronomer. A variation in the Arabic spelling of the name may have caused the lapse: برخس ، ابرخس.

    Hipparchus was born in northwestern Asia Minor, but spent most of his career in Rhodes. He is considered the founder of trigonometry and the one who transformed Greek astronomy from a theoretical into a practical science.

    In writing *Almagest,* Ptolemy exhibits heavy dependence on information garnered by Hipparchus, who shows great dependence on Chaldean (Babylonian) sources (*DSB,* Supplement I, XV, 207–224).

25. Chapter 7, note 10.

26. Aristippus was an advocate of the doctrine that pleasure is the chief end of life.

27. This school was founded by Zeno of Citium about 300 B.C. It teaches that the wise should be free from passion and indifferent to pleasure or pain.

28. Ṣāʿid includes this verse without referring to its famous author. It was written by Abū Nawās, who lived in Baghdad and died around A.D. 810/A.H. 194.

9. *Science of the Romans*

1. The Arabs used the word "Rūm" to designate not only the Romans, but Christians in general. Thus the Greek Orthodox are still called Rūm Orthodox.

2. It is interesting to note that the Arabs assumed that "Ocean" was a proper name, while the Europeans assumed that the word "Sahara" [desert] was the proper name of the African desert.
3. Shaykhū (1912:34) and Blachère (1935:77) have "Almania" in both places, while Bū-ʿAlwan (1985:97) has "Amaynah" in both places.
4. In several copies of the manuscript, the word "Persia" appears instead of Baṣra. But this is an error. Al-Khalil ibn Aḥmad never lived in Persia: he lived in Baṣra and died there in the year A.D. 883/A.H. 269. Al-Khalil was a well-known lexicographer and grammarian. Ḥunayn ibn Isḥaq is considered one of the foremost Arab scholars of his time and an indefatigable translator of medical texts from Syriac and Greek into Arabic (see chapter 12).
5. It is obvious that most of the scientists and philosophers mentioned in this chapter were neither Romans nor Greeks. Ṣāʿid, like other Arabs of his era, was using the word "Rūm" to indicate Christians in particular and non-Muslims in general. The known world was then divided into an Arab Muslim Empire and a Rūm Christian Empire. Most of these scholars were Christian Arabs working in Baghdad, where the government provided them with support and encouragement. Furthermore, we see no justification for including Qusṭā ibn Luka with the Greek thinkers instead of in this chapter.

10. *Science in Egypt*

1. The earliest Egyptian pyramids were built in the thirtieth century B.C., by probably the greatest architect of antiquity, Imhotep. Construction on the Great Pyramid of Gizah began around 2885 B.C. The completion of this royal tomb required between 100,000 and 200,000 men working for about twenty years. It covers about 60,000 square meters and contains some 2,300,000 blocks of limestone whose total weight is about 60 million tons.
2. The bracketed section is missing from several copies of the manuscript.
3. The ancient city of Rashid or al-Rashid, also known as Rosetta, is located on the Mediterranean just east of Alexandria.
4. This myth could have had its origin in prehistoric Egypt, when, after the last Ice Age, a persistent decrease in rainfall drove animals and humans into the Nile gorge. In this narrow enclosure, people and beasts had to compete for food and water, and some animals could have been driven off into the Egyptian desert plateau. But it should also be mentioned that this close contact between animals and humans gradually resulted in the domestication of animals and the invention of the first agricultural machine, the ox-drawn plow. Thus, for the first time, humans began to use power other than their own (Breasted 1936:53–56).
5. This sentence must have been added to the text at a later date; al-Waṣyfī (Ibrahim ibn Waṣyf Shah) died in A.D. 1202/A.H. 598, some 130 years after the death of Ṣāʿid (Kajalah 1959:I, 125).
6. See chapter 7 on science in Chaldea and its notes.
7. Excavations in the Nile Valley reveal that the Hymns of Praise to the sun-god (Ra) were authored by Ikhnaton around 1300 B.C. They were found on the walls of the tomb-chapels in Amarna; in verse, Ikhnaton

describes his god as a kind and loving father and as the creator of all living things. Some passages in Ikhnaton's hymns are similar to those in the Psalms:

From Ikhnaton's hymn	*From the 104th Psalm*
How manifold are thy works! They are hidden before men. O sole God, beside whom there is no other. Thou didst create the earth according to thy will.	O Lord, how manifold are thy works! in wisdom hast thou made them all: the earth is full of thy riches. So is this great and wide sea,

(Breasted 1936: 231)

8. Proclus studied in Alexandria and then lived in Athens. He was an advocate of Neoplatonism.
9. Theon's main work is his commentary on Ptolemy's *Almagest.*
10. Rūshum lived in Egypt prior to the advent of Islam. He was an able chemist (al-Qifṭī 1903: 186).
11. The ruins of Egypt reveal that the Egyptian contribution to human knowledge is momentous. In 4236 B.C. they formulated a most practical calendar, according to which the year was divided into twelve thirty-day months and five year-end feast days. With some alterations, this is our present-day calendar. The next known improvement on the measurement of the length of the solar year came some 3,740 years later. The Egyptian calendar has been acclaimed as the only intelligent calendar in human history and was used by the Hellenistic astronomers and their successors until Copernicus. The Egyptians calculated the areas and volumes of various geometric shapes; some 1,500 years before the Greeks, they formulated the expression $[8/9(2R)]^2$ for the area of the circle. The value of π calculated from this expression is 3.16049, which is only 0.6 percent larger than the present-day best-known value. About 200 B.C., the Egyptian astronomer Aristarchus devised the first heliocentric model of the universe, which did not gain popularity due to the dominance of Aristotle's geocentric model, but was reestablished in A.D. 1514/A.H. 920 by Copernicus; at about the same time, another Egyptian, Eratosthenes, measured the size of the earth and determined its diameter to be 12,640 kilometers. This is less than 0.1 percent smaller than the best-known modern value. It is quite possible that this high accuracy is due more to luck than to ability.

11. The Arabs: General Information

1. Al-Jāhiliyah is the period that preceded Islam. The word means a period of ignorance, indicating that the people of that period did not know about Islam. Blachère substitutes paganism for Jāhiliyah (1935: 88).
2. Bū-ʿAlwan writes "Abū Muʿin, founder of Africa" (1985: 112).
3. Abū Tammām is a highly respected Arab poet. He died around A.D. 846/A.H. 231.
4. This sentence seems to contradict the previous paragraph; but Ṣāʿid could be indicating that the kings of Ḥimyar did not encourage observational astronomy.

5. The bracketed section is missing in Blachère (1935:91) and in Shaykhū (1912:43).
6. *Ḥiys* is a food that the Arabs used to make by combining dates, butter, and curds.
7. Qur'an, al-Zumar [Throngs], XXXIX,3.
8. Here Blachère adds a paragraph about the Arabs acquiring scientific information from the Persians (1935:93).
9. Al-Haytham ibn ʿAdy al-Ṭaʾy was a historian and a man of literature. He flourished in Kūfah, Iraq, and died c. A.D. 822/c. A.H. 206 (Bū-ʿAlwan 1985:119).
10. Aḥmad ibn Dāwūd al-Dynūrī [al-Dīnawarī] (d. A.D. 895/A.H. 282), also known as Abū Ḥanifah, is the author of some thirteen volumes on a variety of topics: botany, history, poetry, theology, language, algebra, and a study of Indian mathematics (*DSB* I, 350; IX, 614; XI, 246).
11. Eilat is a city on the Gulf of ʿAqaba; according to Arab geographers, this is where al-Sham begins. ʿAdhyb is a locality not far from Baṣra.
12. Al-Anṣār means "the supporters." These were the clans who supported the Prophet upon his arrival in Yathrib [Medina].
13. The Prophet was born about A.D. 570/B.H. 54 and died in A.D. 632/A.H. 10.
14. These became known as *al-khulafāʾ al-rāshidūn* [the rightly guided caliphs]. The word *khalifah* means "successor" and implies that the caliph rules by the laws of the Qur'an and the practice of the Prophet.
15. Qur'an, Āl ʿImrān [the family of ʿImrān; used in the Qur'an as a generic name for all the Hebrew prophets], III, 140.
16. Al-Ḥāryth ibn Kaldah (d. A.D. 671/A.H. 51) was a well-known physician, originally from Ṭāʾif. He studied medicine in Persia. The Prophet ordered anyone with an illness to visit him. He authored the book *A Dialogue in Medicine.*
17. Al-Munṣūr reigned as caliph A.D. 754–775/A.H. 136–158.
18. This is probably in reference to the popular use of young Turks as male prostitutes or to the deterioration of the political power of the caliphs of Baghdad at the hands of the Turks. Al-Muʿtaṣim, who became caliph upon the death of his brother al-Maʾmūn in A.D. 833/A.H. 218, was not able to rely on the loyalty of his own army and found it necessary to recruit an army of young Turks from the provinces of Turkestan and Transoxania. In less than thirty years, the Turks were able to dominate the political scene in Baghdad.
19. This sentence is not very clear; it could mean that science was on the decline until it made a comeback during the present time.

12. Science in the Arab Orient

1. Bū-ʿAlwan (1985) does not mention *al-Durrah al-Yatimah,* and Blachère writes "prince" instead of "authority" (1935:101).
2. This astronomer died around A.D. 777/A.H. 160 (al-Qifṭī 1903:270).
3. Ibn al-Ādamī died in the early part of the tenth century, before completing *Naẓm al-ʿIqd,* which was later completed by his student al-Qāsim ibn Muḥammad ibn Hāshim al-ʿAlawī and published around A.D. 920/A.H. 307 (al-Qifṭī 1903:282); Ṣāʿid states that it was pub-

lished in A.D. 949/A.H. 337. The title of the book is given as *Niẓām al-ʿIqd* by Shaykhū (1912 : 58) and by Blachère, who translates it to mean "the thread of the necklace" (1935 : 102).

4. Al-Khuwarizmi, the outstanding mathematician of the Arabs, wrote the famous book titled *Kitāb al-Jabr wa al-Muqabalah.* The word "algebra" is derived from the second word in this title, *al-jabr,* which means "bone-setting." The author chose the word *al-jabr* to provide a graphic description of one of the two operations for the solution of quadratic equations.
5. This sentence is flawed: Shamāsiyah is near the city of Baghdad and not Damascus. Al-Qifṭī reports that observations were carried out in Shamāsiyah and on Mount Qāsiūn near Damascus, al-Sham (1903 : 357).
6. Yaḥyā ibn Abū Munṣūr was a Persian, a member of the Munajjimīn [astronomers] family. He entered the court of al-Maʾmūn, who encouraged him to accept Islam. Yaḥyā is the author of several treatises on astronomy. He died in A.D. 1219/A.H. 616 (al-Qifṭī 1903 : 234).
7. The genealogical listing presented here and in several other parts of the manuscript is probably due to the emphasis the Arabs put on the importance of heredity; the Arabs call their world-famous horse *aṣil* [having good roots]; to compliment a good man, they say *ibn aṣil* and a good woman *bint aṣil;* to put someone down, they say *balla aṣil* [having no roots].
8. For more about al-Aʿsha, see *EI,* I, 48.
9. These four poems are reproduced by Ibrahim Jizzinī (1969 : 86, 151, 202, 206).
10. This is the name given to the Arabs who inhabited the northern regions of Arabia, because they are the descendants of Maʿd ibn ʿAdnān.
11. This phrase is missing in Bū-ʿAlwan's version (1985 : 136). Manichaeism originated in Persia in the third century A.D., teaching the release of the spirit from matter through asceticism.
12. This author served Caliph al-Muʿtaḍiḍ. Blachère gives his name as Aḥmad ibn al-Ṭayyib al-Sarahsi (1935 : 106).
13. Al-Razi wrote 184 works, including a compendium of his experiments, observations, and diagnoses with the title *al-Ḥāwi* [The All-Encompassing]. He was also a physician who wrote books that show common sense, for example: *The Reason Why Some Persons and the Common People Leave a Physician Even If He Is Clever* and *A Clever Physician Does Not Have the Power to Cure All Diseases, for That Is Not within the Realm of Possibility.*
14. Caliph al-Muqtadir ruled in Baghdad A.D. 908–932/A.H. 295–320.
15. Although Ṣāʿid and Bū-ʿAlwan (1985 : 139) give the title of this work as *al-Syrat al-Fāḍilat,* the actual title of this famous book is *al-Madinat al-Fāḍilat* [The Model City] (Munk 1927 : 313; Blachère 1935 : 109).
16. When Baghdad lost its political dominance around A.D. 860/A.H. 245, the Hamdanids, a purely Arab dynasty, ruled over Syria and patronized philosophers and poets such as al-Fārābī and al-Mutanabbi.
17. This table is also known as *Zij al-Shāh* [Astronomical Tables of the King] (*EI,* 1st ed., I, 506).

18. The astrolabe is a compact instrument used to locate the positions of celestial bodies. It was later replaced by the sextant.
19. Mūsā ibn Shākir was a bandit in his youth. Later he studied astrology and served in al-Ma'mūn's court. He died in A.D. 816/A.H. 200 (ibn al-ʿIbri 1890:152).
20. This statement is not very clear; it implies the existence of a book by that title. According to Blachère (1935:111), neither al-Qifṭī (1903) nor ibn al-Nadīm (1871–1872) mentions such a title. The word *kitāb* here means a letter probably written by Abū Maʿshar and sent to Shādhān.
21. Al-Battānī was born in Ḥarrān, northwestern Mesopotamia, hence the name Ḥarranī, but lived most of his life and carried out his famous astronomical observations in the city of Raqqa, situated on the left bank of the Euphrates. Among his best-known works are *Kitāb al-Zij* [Opus astronomium] and *Kitāb Maṭāliʿ al-Burūj* [Book of the Ascension of the Signs of the Zodiac].
22. Caliph al-Muʿtamid ʿAla al-Allah reigned A.D. 870–892/A.H. 256–278. Al-Battānī began his observations in A.D. 877/A.H. 263, the eighth year in the reign of al-Muʿtamid (Blachère 1935:111).
23. Al-Ḥasan ibn al-Ṣabbāḥ ibn ʿAli al-Ismāʿīli was born in Mru, Persia, around A.D. 1037/A.H. 428 and studied in Nishapur under the tutelage of Niẓām al-Malik and ʿUmar al-Khayyam. He was a very able geometer, mathematician, and astronomer. Later on he resided with a band of *ḥashashin* [assassins, addicted to hashish] in Qalʿat al-Mawt [Fortress of Death]. Ibn al-Ṣabbaḥ died in A.D. 1124/A.H. 518.
24. Al-Tanukhī was both an astronomer and an astrologer; he was known for his travels in quest of knowledge (al-Qifṭī 1903:208).
25. Blachère translates this sentence to mean "touching on eternity or the creation of the universe" (1935:115).
26. Blachère translates the title of the book *Kitāb al-Qiwa* to mean "Book of Virtues" (1935:116).
27. Medieval optics, in Christendom and in Islam, was dominated by *Kitāb al-Manāẓir* of ibn al-Haytham [Opticae thesaurus of al-Hazen] (Sarton 1927–1953:II, part I, 23).
28. Abū Zayd ʿAbd al-Raḥmān was born in Córdoba and became the *qāḍi* of Toledo, Tortosa, and Denia. He died in A.D. 1080/A.H. 472.
29. The name of this scholar appears in al-Qifṭī (1903:327) as Misha ibn ʿIbri al-Yahudi [the Jew] and not al-Hindi [the Indian].
30. Abū Qamāsh is the name of a governor of Egypt during the reign of Harūn al-Rashid. Ibn al-Nadīm (1871–1872) mentions that this book was translated from the Syriac and the Indian by Isḥaq al-Hāshmī.
31. Blachère translates the nickname of this physician to mean "Thundering Poison" (1935:118).
32. It is interesting to note that Ṣāʿid and probably the author used the Greek form of the word for the title of this book.
33. Bū-ʿAlwan gives the name of this scholar as al-Ṣufi (1985:152). For more information about al-Kūfi, see *EI,* I, 1015. (Jābir ibn Ḥiyān, known in the West as Geber, died in A.D. 776/A.H. 159.)

34. Al-Muaṣawifyn and Sufism are derived from the Arabic word *ṣūf* [wool], designating a school of thought whose disciples neglect material wealth in favor of spiritual enrichment.
35. Ibn Rabban al-Ṭabarī was born in Mru, near present-day Tehran. At the age of ten, he moved with his father to Ṭabaristan—hence the name Ṭabarī. He was summoned to Baghdad in A.D. 840/A.H. 225. His best-known works, *Firdaws al-Ḥikmah* [Garden of Wisdom] and *al-Dīn wa'd-Dawlah* [Religion and Government], were written A.D. 850–855/A.H. 235–240. Because of its scope and comprehensiveness, *Firdaws al-Ḥikmah* is viewed as a medical encyclopedia. His father was probably a teacher—hence the name Rabban, meaning teacher in Syriac, not from the Arabic meaning, the captain of a ship (*DSB*, XIII, 230).
36. This statement is open to question: al-Razi was born in Rayy, Persia, around A.D. 854/A.H. 239 and was only a youngster when al-Ṭabarī died in A.D. 861/A.H. 246.
37. Al-Qifṭī (1903:232) mentions the title of this book as *Kitāb al-Makki* not *al-Malakī.*
38. Al-Daylamī gained power during the Abbāsid era and became second only to the caliph. During the Islamic period, he was the first ruler to be given the title "king of kings." He encouraged science and scientists and died around A.D. 983/A.H. 372.
39. Ṣāʿid, all through this manuscript, fails to mention scientific organizations. It should be stated that the Abbāsid caliphs established a great center for research named Bayt al-Ḥikmah [House of Wisdom] where a remarkable assemblage of scholars and translators worked. Best known among them are Ḥunayn ibn Isḥaq and his son Isḥaq ibn Ḥunayn. The father, known to the West as Joanitius, translated into Arabic the entire Greek canon of medical works, including the Hippocratic oath, wrote some thirty original treatises on various medical topics, and became the director of Bayt al-Ḥikmah. The son was an accomplished mathematician and geometer who with the help of Thābit ibn Qurrah prepared a critical edition of Euclid's *Elements.* There was also Mūsā ibn Shākir and his three sons: Muḥammad, Aḥmad, and al-Ḥasan, known collectively as Banū Mūsā (see "Banū Mūsā," *DSB*, I, 443). In addition to sponsoring the translation of Greek works, they wrote many original treatises in geometry, algebra, physics, and celestial mechanics; while on a trip to Byzantium, Muḥammad met Thābit ibn Qurrah, one of the greatest of the ninth-century scholars, whose knowledge of Arabic, Greek, and Syriac was unrivaled. Muḥammad made Thābit relinquish his occupation of money changer, brought him to Baghdad, and introduced him to the caliph. In addition to his enormous involvement with translation, Thābit contributed over seventy original works in mathematics, astronomy, physics, music, medicine, philosophy, and the construction of scientific equipment. Other Arab scholars who worked in Bayt al-Ḥikmah include al-Khuwarizmi, al-Razi, ibn Māsawayh, and al-Kindi. It was probably in Bayt al-Ḥikmah that Arab and Indian mathematicians cooperated to produce both the eastern and the western Arabic numerals.

13. Science in al-Andalus

1. Abū ʿUbayda Muslim ibn Aḥmad ibn Abū ʿUbayda al-Laythī (Ṣāḥib al-Qiblah) was a traditionalist. His name appears as al-Balansi in Shaykhū's work (1912:64) and in Blachère's translation (1935:123).
2. Al-Qiblah is the direction in which a Muslim faces, toward Makkah, while performing prayers. "We see your face in the heavens, and we provide you with a Qiblah for guidance, turn your face toward al-Masjid al-Ḥarām, wherever you are turn your face toward it . . ." (Quʾran, al-Baqrah [The Cow], II, 144).
3. ʿAbd al-ʿAziz was a traditionalist who died in Makkah in A.D. 899/A.H. 286.
4. Ismāʿīl ibn Yaḥyā ibn Ismāʿīl Abū Ibrahim al-Māznī was a Shafiʿi scholar who died in Egypt in A.D. 877/A.H. 263 (al-Shyrāzī 1970:97).
5. ʿAbd Rabbih was an Arab Andalusian author and poet. He died in A.D. 940/A.H. 328 (*EI*, II, 375).
6. This is in reference to theological doctrines like *irja* and *iʿtizal.*
7. This is in reference to the changing of the Qiblah from Jerusalem to Makkah.
8. The Arabic word *sahl,* which appears in this verse, has two meanings: "plain" or "simple." Such play on words is commonly used in Arabic poetry.
9. These verses were composed during the later part of the ninth century and the opinion they express has historical merit. Several sources present variations of these verses and their authenticity is in some doubt.
10. Ibn Mūsā died in A.D. 920/A.H. 307. Blachère gives his name as al-Afsin (1935:124) Qurayshite [of Quraysh], the family of the Prophet. Blachère states that Abū ʿUbayda died in A.D. 888/A.H. 274 instead of A.D. 908/A.H. 295 (1935:124).
11. Al-ʾAmir al-Ḥakam al-Mustanṣir bi-Allah ibn ʿAbd al-Raḥmān al-Nāṣir li-Dīn Allah is an Umayyad caliph, who died in September, A.D. 976/A.H. 365 (*EI*, II, 237).
12. Al-Majriṭi wrote a number of works on mathematics and astronomy, elaborated on the Arabic translation of Ptolemy's *Almagest,* and adapted the astronomical table of al-Khuwarizmi to the longitude of Córdoba and to the Hijrah calendar. Of the works attributed to al-Majriṭi, *Ghāyat al-Ḥakīm* [Aim of the Wise] was translated into Spanish in A.D. 1256/A.H. 654 and was widely distributed throughout Europe under the title *Picatrix,* a corruption of Hippocrates. "In general, it may be assumed that the magical and chemical works attributed to al-Majriṭi are spurious, especially since ibn Ṣāʿid does not refer to them in Ṭabaqāt." Al-Majriṭi died in Córdoba, c. A.D. 1007/c. A.H. 397 (*DSB*, IX, 39–42).
13. Related to the seven lectures of the Qurʾan (Blachère 1935:128).
14. The name of this famous scholar appears as al-Marjiṭ and al-Marjiṭi in Bū-ʿAlwan (1985:166, 167, 168, 193); and as al-Marhit in Shaykhū's work (1912:67, 68, 69).
15. For more information about this famous astronomer, see chapter 12.

16. Abū al-Qāsim ʾAṣbagh ibn Muḥammad ibn al-Samḥ flourished in Granada and died in A.D. 1035/A.H. 426. For more about this scholar, see Sarton (1927–1953:I, 715).
17. Note the similarity between the title of this book and al-Majriṭi's book *Thimār ʿIlm al-ʿAdad.*
18. Al-Ṣanhājī ruled in Granada, A.D. 1019–1038/A.H. 409–429, during the decline of the Umayyad dynasty, which flourished in Córdoba until A.D. 1031/A.H. 422.
19. For more information about the astrolabe and the history of its use by Muslim astronomers, see King (1979:245).
20. Al-ʿĀmirī, after the fall of the Umayyad dynasty, carved for himself a princedom that stretched over the province of Denia and the Balearic Islands (*EI,* III, 666).
21. Al-Jazirat is the name the Arabs bestowed on the plain of northwestern Mesopotamia, where the city of Ḥarrān is located.
22. Ikhwān al-Ṣafā was an Arabic school of philosophy. The Arabic word *al-ṣafā* implies a state of purity and happiness (see, for example, *EI,* II, 487).
23. For more information about this scholar and his physician grandfather, Ḥasday ibn Isḥaq, see chapter 14.
24. Abū Muslim ʿUmar is a member of a family known for its many accomplished scholars. He is a descendant of an ancestor named Khaldūn, who settled in Spain in the eighth century. The family gave up its patrician home in Seville in A.D. 1248/A.H. 646, before the Christian conquest of this city, and moved to North Africa, where its members continued their scholarly pursuits. Their most illustrious scholar is ʿAbd al-Raḥmān ibn Khaldūn, the noted historian, political scientist, and sociologist, who flourished in North Africa in the second half of the fourteenth century. ʿAbd al-Raḥmān completed his famous *al-Muqaddimah* [The Introduction] in A.D. 1377/A.H. 779 and died in Cairo, Egypt, in March, A.D. 1406/A.H. 809.
25. Blachère gives the name of this astronomer as al-Kursai and al-Amtas al-Marwānī, as if there were two persons, and omits the paragraph about al-Marwānī (1935:133–134). Bū-ʿAlwan combines ibn Shahr and al-Qarshī al-Afṭas al-Marwānī as if they were one name (1985:173).
26. Al-ʿĀmirī governed Almería, as its prince, after the fall of the Umayyad dynasty (see note 20 above).
27. Al-Ẓāfir ruled the princedom of Toledo, A.D. 1035–1037/A.H. 426–428.
28. This date is given by Shaykhū (1912:73) and by Blachère (1935:135) as A.D. 1015/A.H. 405, which must be an error: ibn al-Layth was a student of ibn Barghūt, who died in A.D. 1053/A.H. 444.
29. In various copies of the manuscript, the name of this prince appeared as al-Salhi, al-Tulayḥi, al-Ṣabḥī, and so forth. But the founder of the Ṣulayḥid dynasty is ʿAli ibn Muḥammad al-Ṣulayḥī, imam of Yemen, who died in A.D. 1080/A.H. 472 (*EI,* IV, 540).
30. Al-Mahdī is the founder of the Fatimid dynasty that ruled Egypt and most of North Africa for some three hundred years beginning in A.D. 1036/A.H. 427.

31. Al-Qā'im bi 'Amr al-Allah ruled in Baghdad as caliph, A.D. 1031–1075/A.H. 422–467 (*EI*, I, 683).
32. Al-Waqshi was Ṣā'id's mentor. See the introductory chapter on Ṣā'id.
33. Blachère presents this information about al-Waqshi in a different order (1935 : 137).
34. Shaykhū gives the name of this scholar as 'Āmir ibn Minh (1912 : 47).
35. Al-Tajibī was also one of Ṣā'id's teachers; see the introductory chapter on Ṣā'id.
36. Al-Zarqāli was Ṣā'id's accomplished student; see the introduction. He died in A.D. 1087/A.H. 480. Bū-'Alwan gives his name as al-Zarqiyal (1985 : 180, 181) and Blachère names him al-Zarkiyal (1935 : 138, 139).
37. This amir governed A.D. 1081–1085/A.H. 473–477 under the name Abū 'Āmir Yūsuf al-Mu'tamin.
38. Toomer (1968, 1970).
39. For more about ibn Ḥazm, see the introduction.
40. Al-Aṣbahānī is the founder of the al-Ẓāhir rite. He died in A.D. 883/A.H. 269. The Arabic word *ẓāhir* means "apparent" or "outstanding." Al-Ẓāhir is a Sufi school of thought that teaches esoteric views (see chapter 8, note 5).
41. For more about al-Ṭabarī, see chapter 12, note 35.
42. Al-Farghānī died around A.D. 980/A.H. 369. Only fragments from his book on the history of al-Ṭabarī have survived.
43. *Al-Ṣilah* means "connection" and not "the present" as stated by Blachère (1935 : 141).
44. *Al-Gharyb al-Muṣannaf* was written by 'Ubayd al-Harawi, who died c. A.D. 837/c. A.H. 222 (*EI*, I, 114); and *Iṣlāḥ al-Manṭiq* was written by ibn al-Sikkit, who died around A.D. 857/A.H. 242 (*EI*, II, 444).
45. This book is a collection of ancient poems put together by Abū Tammām, one of the giants of Arab poetry, who died in A.D. 846/A.H. 231.
46. Bū-'Alwan gives the name of this physician as Aḥmad ibn Iba (1985 : 186).
47. 'Abd al-Raḥmān al-Awsaṭ reigned as caliph A.D. 852–886/A.H. 237–272. Blachère takes al-Awsaṭ to mean "second in name" (1935 : 143).
48. This physician is Yūnus ibn Aḥmad al-Ḥarranī, the father of Aḥmad and 'Umar, two well-known al-Andalus physicians (al-Qifṭī 1903 : 394–395).
49. The complete title of this famous book is *al-'Iqd al-Farid* [The Unique Necklace].
50. Ibn al-Kinānī served al-Nāṣir and al-Mustanṣir bi-Allah. His nephew and namesake was also a known physician. All the copies studied by Bū-'Alwan give the name of these physicians as ibn al-Kinānī, and she reports it as ibn al-Kattānī (1985 : 190, 192).
51. For more information about this physician, see ibn Jaljal (1955 : 80, 81, 112).
52. Ibn Waṣyf was a very able eye doctor who lived in Baghdad toward the middle of the fourth century A.H. (al-Qifṭī 1903 : 436).
53. Al-Mu'ayyad bi-Allah ruled as caliph A.D. 976–1009 and 1010–1013/A.H. 365–399 and 400–403.

54. Al-Sajistānī al-Baghdādī was a celebrated ninth/tenth century logician, who served in the court of ʿAḍad al-Dawlah (al-Qifṭī 1903:282).
55. *Al-taksyr* is the study of fractions and not land surveying as indicated by Blachère (1935:147).
56. Blachère does not mention "al-ʾUsquf," but indicates in a footnote that they were ecclesiastical Christians, although they have Muslim names (1935:149). Bū-ʿAlwan gives the name of this scholar as al-Marjiṭ (1985:193).
57. Blachère omits this sentence (1935:150).
58. Marwān ibn Zuhr is a member of a family of famous scholars and physicians: his son, Abū al-ʿAlāʾ ibn Zuhr, was a physician who served the Murabiṭ dynasty in Spain; his grandson, Abū Marwān ʿAbd al-Malik ibn Zuhr, is the famous physician known in the West as Avenzoar. The ibn Zuhr belong to the Arabian tribe of Iyad. They settled in Spain in the tenth century (*DSB,* XIV, 637).
59. Denia was then governed by Prince Mujāhid, who appointed ibn Zahr to his court and made him one of his closest confidants.
60. Ibn al-Nabbāsh was previously mentioned in the manuscript; Ṣāʿid also included him with the naturalists.
61. Abū Jaʿfar ibn Khamīs ibn ʿĀmir ibn Damj was mentioned earlier as a mathematician and an astronomer.
62. Bū-ʿAlwan, based on a single copy of the manuscript, in the Chester Beatty Library (Dublin), calls this physician al-Darmi (1985:198); both Shaykhū (1912:86) and Blachère (1935:153) mention that he was an eloquent speaker.
63. Al-ʾIstijī was mentioned earlier as a student of philosophy working in Toledo.
64. The East and the West here refer to the contents of chapter 12 and chapter 13, respectively.

14. *Science of Banū Israel*

1. The Hebrew calendar is a luni-solar calendar that is reckoned from the year 3761 B.C., the date traditionally given for the creation. It received its present fixed form from Hillel II (A.D. 330–365) in the year A.D. 359.
2. Each leap year has thirteen lunar months and a total of 383 days (defective), 384 days (regular), or 385 days (perfect, or abundant).
3. The lunar or defective year has 353, 354, or 355 days.
4. The information provided here is in error (see notes 1 and 2 above). The actual difference in the length of the two years as defined in this paragraph is 10 days, 21 hours, and 207.36 minutes.
5. ʿUmar ibn ʿAbd al-ʿAziz was an Umayyad caliph who reigned A.D. 701–720/A.H. 81–101.
6. This physician was mentioned in chapter 12.
7. Al-Mahdī was the founder of the Fatimid dynasty that ruled Egypt and parts of North Africa for some three centuries, beginning early in the tenth century.
8. Abū al-Faḍl Ḥasday is the grandson of Ḥasday ibn Isḥaq, who was mentioned earlier.

9. This is in error. This Jewish theologian is Saʿadiya ibn Yūsuf al-Fayyūmī. He was born in Egypt and died in Mesopotamia in A.D. 942/A.H. 330.
10. In the manuscript, the name of this famous theologian is given as al-Ghazal, as he was known to the Arabs. His Hebrew name is Samual Ha-Levi ibn Nagdila and he served as the vizier of al-ʾAmir Bādīs of Granada (*EI*, I, 567).

Bibliography

Baḥr al-ʿUlum, M. 1967. *Ṭabaqāt al-ʾUmam.* Baghdad, Iraq: al-Najaf, al-Maṭbaʿah al-Ḥaydariyah.

ibn Bashkuwal, Abū al-Qāsim Khalaf ibn ʿAbd al-Malik. 1966. *Kitāb al-Ṣilah fi Tārikh Aʾimat al-Andalus.* Ed. ʿIzat al-ʿAṭṭar al-Ḥusayni. 2 vols. Cairo: Turathna, al-Maktaba al-Andalusīyah, 4–5.

Bell, Richard, and W. W. Montgomery. 1972. *Introduction to the Qurʾan.* Edinburgh: Edinburgh University Press.

Blachère, Régis. 1935. *Kitâb Tabakât al-Umam (Livre des catégories des nations).* 5th ed. Paris: Larose Editeurs.

Breasted, J. H. 1936. *Ancient Times, A History of the Early World.* Boston: Ginn and Company.

Bū-ʿAlwan, Hayat. 1985. *Ṭabaqāt al-ʾUmam.* Beirut: Dār al-Ṭaliʿāt l-Ṭibāʿat wa al-Nashr.

Chaube, R. K. 1969. *India as Told by the Muslims.* Varansi, India: Prithivi Prakashan.

Chejne, A. G. 1974. *Muslim Spain: Its History and Culture.* Minneapolis: University of Minnesota Press.

Dictionary of Scientific Biography. 1970–1980. Ed. Charles Coulton Gillispie. New York: Scribners.

Doniach, N. S. 1962. *The Oxford English-Arabic Dictionary.* Oxford: Clarendon Press.

Encyclopedia of Islam. 2nd ed. 1960–1978. Leyden: E. J. Brill. (1st ed., Leyden: E. J. Brill, 1934.)

Galen. 1952. On the Natural Faculties. In *Great Books of the Western World.* Ed. R. M. Hutchins. Chicago: Encyclopaedia Britannica.

Glassé, Cyril. 1989. *The Concise Encyclopedia of Islam.* San Francisco: Harper and Row.

Haskins, C. H. 1927. *Studies in the History of Medieval Science.* Cambridge: Harvard University Press.

Huart, C. A. 1966. *History of Arabic Literature.* Beirut: Khayat.

———. 1972. *Ancient Persia and Iranian Civilization.* Trans. M. R. Dobie. New York: Barnes and Noble.

al-Ḥūmayrī, ʿAbd al-Munʿim. 1980. *al-Rawḍ al-Muʿṭar.* Ed. ʾIḥsan ʿAbbās. Beirut: Dār al-Thaqāfah.

ibn al-ʿIbri, Abū al-Faraj Yūḥanna al-Malṭī [also known as Bar Hebraeus]. 1890. *Tārikh Mukhtaṣar al-Duwal.* Ed. al-Ṣālihāni. Beirut: Catholic Press.

ibn Jaljal. 1955. *Ṭabaqāt al-Aṭubāʾ wa al-Ḥukamāʾ.* Ed. Fuād al-Sayyid. Cairo: Maṭbaʿat al-Maʿhad al-ʿIlmī al-Farancī Lil ʾĀthār al-Sharqiyah.

Jizzinī, Ibrahim. 1969. *Sharih Diwān al-Aʿsha.* Beirut: Dār al-Kitāb al-ʿArabī.

Kajalah, ʿUmar Raḍa. 1959. *Muʿjam al-Muʾallifyn.* Damascus: Maṭbaʿat al-Taraqqi.

Karpinskii, L. C. 1965. *The History of Arithmetic.* New York: Russell and Russell.

ibn Khaldūn, ʿAbd al-Raḥmān. 1958. *al-Muqaddimah.* Trans. and ed. Franz Rosenthal. Princeton: Princeton University Press.

Khalidī, Ṭ. 1975. *Islamic Historiography.* Albany: State University of New York.

Khan, M. S. 1980. Qādi Ṣāʿid Account of Medieval Arab Astronomy. *Islamic Culture* 54:153.

King, D. A. 1979. On the Early History of the Universal Astrolabe in Islamic Astronomy, and the Origin of the Term "Shakkāzīya" in Medieval Scientific Arabic. *Journal of the History of Arabic Science (JHAS)* 3: 244–257.

King, D. A., and G. Saliba. 1987. *From Deferent to Equant.* New York: Academy of Science.

Lawrence, B. B. 1976. *Shahrastani on Indian Religion.* The Hague: Mouton.

Lichtenstadter, Ilse. 1976. *Introduction to Classical Arabic Literature.* New York: Schocken Books.

Mahdi, Muḥsin. 1957. *Ibn Khaldūn's Philosophy of History.* London: George Allen and Unwin, Ruskin House.

al-Masʿūdī, Abū al-Ḥasan ʿAli ibn al-Ḥusayn ibn ʿAli. 1966–1970. *Murūj al-Dhahab wa Maʿādin al-Jawhar.* Trans. Charles Pellat. 3 vols. Beirut: Lebanese University Press.

Minorsky, V. F. (trans.). 1937. *Ḥudūd al-ʿĀlam: The Regions of the World.* London: Oxford University Press.

Munk, S. 1927. *Mélanges de philosophie juive et arabe.* 2nd ed. 2 vols. Paris: Librairie Philosophique. Vol. 1.

ibn al-Nadīm, Abū al-Faraj Muḥammad ibn Isḥāq Abū Yaʿqūb. 1871–1872. *Kitāb al-Fihrist.* Ed. G. Flügel. 2 vols. Leipzig: F. C. W. Vogel.

Pingree, David. 1968. *The Thousands of Abū Maʿsher.* London: Warburg Institute.

Plesser, Martin. 1956. Der Astronom und Historiker ibn Ṣāʿid al-Andalusī und seine Geschichte der Wissenschaften. *Revista degli Studi Orientali* 31:235.

al-Qifṭī, Abū al-Ḥasan ʿAli ibn Yūsuf. 1903. *Tārikh al-Ḥukamāʾ*. Ed. J. Lippert. Leipzig: Dieterich'sche Verlagsbuchhandlung.

Sachau, E. C. 1964. *Alberuni's India.* 2 vols. New Delhi: S. Chand.

Sarton, George. 1927–1953. *Introduction to the History of Science.* 3 vols. Washington, D.C.: Carnegie Institute.

———. 1954. *Ancient Science and Modern Civilization.* Lincoln: University of Nebraska Press.

Shaykhū [Cheikho], Louis. 1912. *Kitāb Ṭabaqāt al-ʾUmam.* Beirut: Catholic Press.

al-Shyrāzī. 1970. *Ṭabaqāt al-Fuqahāʾ*. Ed. Iḥsān ʿAbbās. Beirut: Dār al-Rāʾid al-ʿArabi.

Toomer, G. J. 1968. A Survey of the Toledan Tables. *Osiris* 15:1–174.

———. 1970. The Solar Theory of Az-Zarqāl. *Centaurus* 14:306–336.

al-Yaʿqūbī, Aḥmad ibn Abū Yaʿqūb. 1883. *Ibn Wadhih qui dicitur al-Jaʿqūbī.* Ed. M. T. Houtsma. Leiden: E. J. Brill.

Index

ibn al-Abbār, xvii
ibn al-ᶜAbbās, ᶜAli. *See* al-Majūsī
Abbāsid, xix, xxi, xxii, 32, 33, 45, 46, 88, 89, 99
ibn ᶜAbd al-ᶜAziz, ᶜUmar, 44
ᶜĀd, 38
Adab al-Nafs, 48
ibn al-Ādamī, al-Ḥusayn, 13, 46, 53, 100
Adelard of Bath, xxv
ᶜAdnān, 4, 38, 43, 44, 97
ibn ᶜAdwī, Abū al-Qāsim, 63
ibn ᶜAffān, ᶜUthmān, 16, 17, 43
ᶜAhd al-Fitnah, 58, 64, 75, 78, 81, 83
ibn Aḥmad, ᶜAbd Allah. *See* al-Sarqasṭi
ibn Aḥmad, Abū Bakr. *See* al-Khayyaṭ
ibn Aḥmad, Abū al-Qāsim, 64
al-Aḥmymī, ibn Ibrahim, 56
Ahwāz, 17
Akkad, 90
Aldebaran, xxiii, 40
Alexander, son of Phillipus, 15, 20, 25–28, 36, 88, 91
Alexander of Aphrodisy, 25, 92
Alexandria, xviii, xxi, 27, 36, 37, 51, 91, 93–95
Alfonso VI, xi
Alfred of Sarashel, xxv
algebra, 89, 96, 97, 99
ibn ᶜAli, Sanad, 47
al-ᶜAllaf al-Baṣrī, Muḥammad, 21
Almagest, 19, 27, 28, 37, 47, 51, 52, 64, 91, 93, 95, 100
Almería, xi, 59, 66, 68, 101
Alpetragius, xxiii
alphabet(ic), xvi, xxiv, 4, 71
al-Aᶜmā, Abū al-Ḥasan, 71
ᶜAmaliqah, 35, 38, 41
Amman, 41
ᶜAmūr, 4
Anaxagoras, 25
Andalusia (al-Andalus), xi, xii, xvii, xxii, 31, 58, 59, 83, 85
Andalusian, xviii, xxiii, xxiv, 83, 100
Ankara, 33
al-Anṣār, 43
al-Anṣāry, ibn Beryal. *See* ᶜAbd al-Bāqi
Anusharwan, 13
apogees, 13, 16, 47
Apollonius of Perga, 26, 27, 92
Arabs, xxi, xxiii, xxv, 6, 9, 10, 40, 42
Archimedes, 27, 93
Aries, 12, 13, 16
Aristarchus of Samos, 26, 95
Aristippus, 29
Aristotle, xiii, xxi, 21, 23, 25, 26, 28–30, 45, 46, 50, 63, 70, 76, 81, 86, 92, 95
Ārjbahd, 13, 88
al-Arkān (The Foundations), 26
Ārkand, 13, 88
Armenian, 4, 20, 31
ibn al-ᶜĀṣ, ᶜUmar, 36
ibn ᶜAsākir, Abū al-Ḥasan, 77
al-aṣbagh, ᶜAbd al-Malik. *See* al-Qarshi
Aṣbahān, 17
al-Aṣbahānī, Dāwūd ibn ᶜAli, 70, 102
Asclepiades, 26
Asrār al-Nujūm . . ., 19
Assyrian, 4

astrolabe, 51, 56, 64, 65, 69, 98, 101
astrology, xxv, 12, 52, 54, 55, 67, 77, 78, 88, 90, 98
astronomy, xxii, xxv, 33, 53, 77, 88, 92
al-Āthār al-ᶜAlawiyah, 24
Atlantic Ocean, 4
Atrahasis, Epic of, 91
ibn al-ᶜAṭṭār, Muḥammad ibn Khyrah, 66
Augustus, 28, 31
Avicebron. *See* ibn Yaḥyā, Sulaymān
Avicenna (ibn Sīnā), xxi, xxiii, 89
ibn al-ᶜAwwām, xxiv
ᶜAbd al-ᶜAziz, ibn Jawshān, 69

al-Baᶜalbeki, Qusṭā ibn Luka, 25, 92
Babel, 89–91
Babylon(ia), xxi, 16, 18, 19, 20, 91
Babylonian, 4, 18, 19, 86, 90, 91, 93
ibn Bādīs, al-Muᶜizz, xxi
ibn Badr, ᶜAbd ar-Raḥmān. *See* al-Iqlidi
ibn Badr, Ismāᶜīl, 63
Baghdad, xix, xxiii, 33, 34, 49, 50, 56, 61, 68, 74, 80, 84, 88, 92, 94, 96, 97, 99, 102
al-Baghdādī, ibn ᶜAbd Allah. *See* Ḥabash
al-Baghiyat, 56
ibn al-Baghūnish, Abū ᶜUthmān, 63, 74, 75, 77
ibn Bahrām, Abū Sulaymān, 74
al-Bajjānī, Abū ᶜAbd Allah, 75, 77
al-Bajjānī, ibn an-Nabbāsh, 71, 77
al-Bajjānī, ibn Maryn, 75
Bakhtishūᶜ, Jibriyl, 33
Abū Bakr al-Ṣiddyq, 43
Balkan, xix
al-Balkhi. *See* Abū Maᶜshar
ᶜAbd al-Bāqi, xvi
barbarians, xxii
ibn Barghūt, Muḥammad ibn ᶜUmar, 66, 67, 101
al-Barmakī, ibn al-Jahm, 55
al-bāṭiniyah, 21, 56, 92
al-Battānī, Abū Jaᶜfar, 52, 53, 64, 98
al-Battānī, ibn Jābir, 28
ibn Baṭṭuṭah, xxiv
Bayt al-Ḥikmah, 99
ibn al-Bayṭar, xxiv
Beirut, xv, 84, 105–107
Berbers, 4, 6, 8
al-Birūnī, 87, 88
Bistāsif, 17, 89
al-Bitruji. *See* Alpetragius
Black Sea, 4, 20, 86
Boga [Bajah], 8, 86
Book of Fevers, 33
Book of Leprosy, 33
Book of Memoirs, 33
Book of Nutrition, 33
Book of Pandects, 33
Book of Perfection, 33
Book of Proof, 33
Book of Purgatives, 33
Book of Roger, The, xxiv
Book of Vision, 33
Bosphorus, 31
botany, xxiv, 96
Brahma, xviii
Brahmagupta, 13
Brahmin, 12, 30, 87
Budasaf al-Mushriqi, 16, 88
Buddha, 25
Bulgarians, 6, 7, 21
Burzuwaih, 13
Bustān al-Ḥikmah, 80

calendar, xvii, xxvi, 59, 62, 64, 81, 95, 100, 103
Canaᶜan, 4, 18
Canopus, 40
Chaldea, xxi
Chaldeans, xix, 4, 6, 18, 19, 86, 87, 90, 91, 94
Charmides, 27
chemistry, 36, 37, 44, 56, 63, 77, 90
chess, 14
China, xxv, 4, 7, 11, 16, 20, 35
Chinese, xix, 6, 7, 9, 11, 88
Christian(ity), xii, 17, 31–33, 35,

40, 41, 58, 71, 72, 84, 92–94, 101, 103
Chrysippus, 29
Cleopatra, 28
conjunctions, 54
Constantinople, 31, 32
Córdoba, xi, xxiii, 21, 58–60, 62, 63, 65–67, 70–75, 98, 100, 101
culture, xviii, xxiii, 80, 85, 105, 106
Cyrene, 29

al-Dākhil, ʿAbd ar-Raḥmān, 55, 59, 72
ibn Damj, Abū Jaʿfar. *See* Domingo
al-Damynah, ibn Dhi ibn ʿUmrū, 54
Daniel of Morley, xxv
Darius, 15, 20, 25, 91
Dark Ages, xxii
David, 21, 40
al-Daylamī, Rukn ad-Dawlah, 57, 99
Democritus, 25
ibn al-Dhahabī, Abū Muḥammad, 77
al-Din wa'd-Dawlah, 99
Diogenes, 29
Domingo, 68
al-Durrah al-Yatimah, 46
dustboard arithmetic, 14, 88
al-Dynūrī, Aḥmad ibn Dāwūd. *See* Abū Ḥanifah

eclipse, xxii, 46, 90
Egypt, xvii–xix, xxi, 4, 6, 18, 21, 27–29, 32, 35–37, 44, 55–57, 60, 61, 67, 74, 76, 82, 84, 85, 87, 90–92, 94, 95, 98, 100, 101, 103, 104
Empedocles, xxi, 21, 22, 92
equinox, 27, 28, 90
Ethiopia(n), 6, 8, 12
Euclid, 26, 27, 45, 52, 63, 64, 93, 99
Euctemon, 27
Europe (Europian), xvii, xxii–xxv, 86, 91, 93, 94, 100

Fādin, 22
al-Faḍl ibn Sahl, 52, 55, 70
Fam al-Dhahab, 48
al-Fārābī, Abū Naṣr Muḥammad, xxii, 49–51, 97
al-Faraghānī, Abū Muḥammad, 70
al-farā'iḍ, 62, 66
al-Farāyḍī, Abū Ghālib, 62
al-Farghānī, Aḥmad ibn Kathīr, 51
Farkhān (Farrukhān), ʿUmar, 51, 52, 55
al-Farq Bayn al-Ḥayawān al-Nātiq wa al-Ṣāmit, 25
al-Farq Bayn al-Nafs wa al-Ruḥ, 25
Faṭūn, 27
ibn al-Fawwāl, Manāḥym, 81
ibn Fayrūz, ibn Qibād, 13
al-Fayyūmī, ibn Yaʿqūb, 82, 104
al-Fazārī, Muḥammad, 13, 46, 55, 88
Fi Ithhāt al-Nubuwah, 48
Fi ma Baʿd al-Ṭabyʿah, 24, 48
Firdaws al-Ḥikmah, 56, 99
ibn Firnas, ʿAbbas, xxiii
Flood, the, 18, 19, 35, 36, 42, 53, 90
France, 31, 59
French, xv, 84, 87
al-Furs, 3, 4, 11
al-Fuṣūl, 26

Galen, 26, 33, 37, 45, 71, 72, 74, 76, 77, 92
Galicians, 8, 32
Genesis, 36, 89
geography, xvii, xxii, xxiv
geometry, xvii, xxii, 12, 21, 25, 26, 33, 34, 47, 49, 51, 52, 63–68, 74–77, 81, 93, 99
Ghana (Gana), 6, 8, 86
al-Gharyb al-Muṣannaf, 71
Ghassān, 40, 41
al-Ghazal, 82, 104
Gilgamesh, Epic of, 91
God, unity of, 12
grammar, 28, 60, 61, 66–68, 71, 73
Granada, 59, 65, 82, 101
Greece, xxiii, xxi, 20, 21

Greek, 4, 6, 19–23, 25–29, 31–33, 35, 41, 55, 87, 90, 92–95, 98, 99

Ḥabash, 13, 51
ibn Ḥabib, Sahl ibn Bushr, 80
ḥadīth, xii, 61
Hadrian, 27
ibn Ḥafṣūn, Aḥmad, 73
al-Ḥājib Abū ʿĀmir, xxii, 69
al-Ḥakam, ʿAbd al-Raḥman, xxii, 61–63, 74, 80
ibn Ḥakam, Masiḥ, 33
al-Ḥakim, 61
Ha-Levi, Samual, 82, 104
Ham, 18
al-Hamdānī, Abū Muḥammad al-Ḥasan, 54, 55
al-Ḥammar, Abū ʿUthmān, 63
Hammurabi, 91
Ḥanif, 17
Abū Ḥanifah, 42, 96
Ḥarrān, 72
al-Ḥarranī, ibn Sinān, 52
ibn Ḥasday, Abū al-Faḍl, 71, 81, 103
Ḥasday ibn Isḥaq, 80
Hāshimite, 44
Haskins, C. H., 85, 105
ibn Ḥātim, al-Faḍl, 28
al-Hawaznī, Abū Isḥaq. *See* al-Ishbilī
al-Ḥāwi, 89
ibn al-Haytham, al-Ḥasan, xviii, 55, 85
ibn Ḥayy, al-Ḥasan. *See* al-Tajibī
ibn Ḥazm, Abū Muḥammad, xxii, xxiv, 67, 69, 70
Hebrews, xix, xxii, 4, 6, 19, 36, 81, 82, 84, 87, 103, 104
Hellenic, 21, 31
Hermes, 19, 36, 90
Ḥijāz, 4
Hijrah, xxvi, 16, 17, 45, 59, 62, 64, 100
ibn Ḥilan, Yūḥanna, 49
Ḥimyar, 39, 40, 41, 54
Hindus, xviii, 87, 88
Hipparchus, 19, 27, 90, 93
Hippocrates, 26, 33, 45, 71, 72, 92, 100
Ḥirah, 41
Ḥisab al-Ghubār, 14, 88
al-Ḥiss wa al-Maḥsus, 24
al-Ḥiyal, 23
ibn Hūd, Abū ʿĀmir, 69, 71
al-Ḥudūd wa al-Rusūm, 80
ibn Ḥusayn, Abū al-Walīd (ibn al-Kinānī), 73, 102

Iblis. *See* Satan
ibn al-ʿIbri, xvii, 92, 106
al-Ibryshim, 72
ibn Idris, Abū Isḥaq. *See* at-Tajibī
al-Idrisi, ibn Idris al-Sharif, xxiv
al-Ihryshim, 72
Iḥṣaʾ al-ʿUlūm, 50
al-ʿilm al-ilāhi. See theology
al-ʿilm al-madanī, 50
ibn ʿImrān, Isḥaq, 56, 80
India(n), xxiii, xxiv, 4, 6, 7, 11, 12–14, 16, 20, 25, 35, 41, 42, 46, 52, 64, 86–88, 98, 99, 105, 106
al-Iqlidi, 63
Iran, xxiii, 65, 89
Iraq, xviii, 3, 4, 17, 39, 42–44, 57, 74, 96
Abū ʿĪsā, Abū Bakr, 63
ibn ʿĪsā, Abū Zayd ʿAbd al-Raḥmān, 55, 69, 98
ibn Isḥaq, Yaḥyā, 72
al-Ishbilī, Abū Marwān, 76
Iṣlāḥal-Manṭiq, 71
Islam(ic), xii, xviii, xxi–xxvi, 7, 14, 17, 25, 32, 34, 38, 43, 44, 47, 48, 52, 55, 58, 60, 61, 70, 85–87, 91, 92, 95–99, 105, 106
Israel, xvi, 4, 18, 41, 42, 79
Istanbul. *See* Constantinople
al-ʾIstijī, Abū Marwān, 69, 78, 103
al-ʾIʿtimād, 56
ibn ʾIyas, Aḥmad, 71

al-Jabalī, ibn ʿAbdūn, 74, 75
al-Jabalī, Muḥammad ibn ʿAbd Allah, 21
al-Jadhāmī. *See* ibn Hūd, Abū ʿĀmir

al-Jāhiliyah, 38, 39, 43, 95
ibn al-Jallab, al-Ḥasan, 67, 68
Jalūlā, 17
Jāmāseb, 16, 88
al-Janyn, 26
al-Jawharī, al-ᶜAbbās, 47, 53
ibn Jawshān, Abū Jaᶜfar, 69
Jazirat al-ᶜArab, 4, 42, 80
al-Jazzar, Aḥmad ibn Ibrahim, 56
Jews, xxii, 17, 41, 79–82, 84, 98, 104
ibn Jināḥ, Marwān, 81
John of Seville, xxv
Judaism, 40
ibn Juljul, Sulaymān, 75
Jupiter, 40, 53, 60, 86
Jurisprudence, 70

Kaᶜbah, 40
ibn Kaldah, al-Ḥāryth, 44, 96
Kalīlah wa Dimnah, 13, 14, 46, 88
Kanka al-Hindi, 14, 88
Kanz al-Muqill, 81
al-Karmānī, Abū al-Ḥakam, 64, 65
al-Kawn wa al-Fasād, 24
ibn Khaldūn, ᶜAbd al-Raḥmān, 101, 106
ibn Khaldūn, Abū Muslim, xviii, 64, 66, 67, 101
Khalifah, Ḥaji, xviii
al-Khalil ibn Aḥmad, 33, 94
ibn Khallikan, xv
ibn Khamīs, Abū Jaᶜfar, 68, 77, 103
Khan, M. S., xviii, 85, 106
Khawarizm, 3
al-Khayyam, ᶜUmar, xxi, 89
ibn al-Khayyaṭ, Abū Bakr, 77
Khorasan, 3, 7, 17, 44, 88
al-Khuṭūṭ, 23
al-Khuwarizmi, Muḥammad ibn Mūsā, xiii, xvii, xxii, 13, 14, 47, 51, 53, 64, 89, 97, 99, 100
al-Kinānī. *See* al-Waqshi
al-Kinānī, ibn Abḥar, 44, 102
ibn al-Kinānī, Muḥammad, 74, 75, 102
Kindah, 38, 40, 48
al-Kindi, Yaᶜqūb ibn Isḥaq, xxii, 25, 26, 33, 42, 48, 49, 92, 99
Kitāb al-Adwiyat al-Mūsahilat, 33
Kitāb al-Aghdhiyat, 33
Kitāb al-Aqālīm, 53
Kitāb al-ᶜArḍ, 19
Kitāb al-Baṣirah, 33
Kitāb al-Burhān, 33
Kitāb al-Duwal wa al-Milal, 53
Kitāb al Fasḍ, 33, 56
Kitāb al-Ḥamāsah, 71
Kitāb al-Ḥayawān, 24
Kitāb al-Ḥummayat, 33
Kitāb al-ʾIklil, xxi, 39, 54
Kitāb al-ᶜIqd, 72, 102
Kitāb al-Istiqṣāt, 80
Kitāb al-Jabr wa al-Muqabalah, 97
Kitāb al-Jidham, 33
Kitāb al-Kamāl, 33
Kitāb al-Madkhal al-Kabīr, 53
Kitāb al-Maᶜīdat, 33
Kitāb al-Makki, 99
Kitāb al-Malāḥim, 53
Kitāb al-Malakī, 57, 99
Kitāb al-Malankhulya, 56
Kitāb al-Manāẓir, 98
Kitāb al-Masāʾil wa al-Ikhtibārāt, 80
Kitāb Maṭāliᶜal-Burūj, 98
Kitāb al-Mawālyd wa Taḥāwylihā, 80
Kitāb al-Mūḥakkam, 71
Kitāb al-Nabaḍ, 56
Kitāb al-Nabāt, 24
Kitāb al-Nafs, 24
Kitāb Naẓm al-ᶜIqd, 53
Kitāb Nuzhat al-Nafs, 56
Kitāb al-Qānūn, 37
Kitāb al-Qiranāt, 53
Kitāb al-Qiwa, 54
Kitāb al-Samāʾ wa al-ᶜĀlam, 24, 82
Kitāb al-Shabāb wa al-Haram, 24
Kitāb al-Ṣiḥat wa al-Ṣuqm, 24
Kitāb al-Ṭabāʾiᶜ, 53
Kitāb Taḥawil Siny al-ᶜĀlam, 80
Kiāb al-Taqryb li-Ḥudūd al-Manṭiq, 70
Kitāb al-Taᶜrif bi Ṣaḥiḥ- al-Tārikh, 57

Kitāb al-Ṭūl, 19
Kitāb al-ʾUlūf, xix, 14, 53
Kitāb Zād al-Musāfir, 56
Kitāb al-Zij, 98
al-Kiyān, 81
al-Kūfi, Jābir ibn Ḥiyān, 56, 98

al-Lakhmī, Abū al-Muṭarraf, 75, 76
language, xii, xiii, xvi, xvii, xix, xxv, 3–5, 13, 20, 21, 28, 31, 37, 41, 44, 46, 58, 60, 61, 63, 66–69, 71, 73, 81, 84, 87, 89, 96
Latin, xii, xxi–xxv, 31, 32, 84, 89
law, xii, xxii, 8, 36, 37, 43, 44, 50, 53, 58, 60–62, 66, 70, 79–82, 87, 91, 92, 96
Lawrence, B. B., xviii, 85, 106
ibn al-Layth, Muḥammad ibn Aḥmad, 67, 101
al-Laythī, Abū ʿUbayda, 60, 100
Leonardo da Vinci, xxiii
logic, 21, 24, 25, 28, 33, 44, 46, 48, 49, 50, 55, 61, 63, 65, 68–70, 73–77, 80, 81, 100
Louvain, xxiv
ibn Luka, Qusṭā (al-Baʾalbeki), 25, 34, 92, 94
Lukman, 21, 91

Macedon, 25, 28
Maʿd, 48, 97
Ahl Madar, 39
al-Madkhal ila al-Handasa, 25, 64
al-Madkhal ila ʿIlm Hayʾat al-Aflāk wa Ḥarakāt al-Nujūm, 25, 51
al-Maghrib, 4, 56
Magus (Majus), 17, 84
al-Mahdī, 48, 68, 80, 101, 103
Mahdi, M., xviii, 85, 106
al-Majriṭi, Muslamah, xxiii, 62–64, 75, 78, 100, 101
al-Majūsī, ʿAli ibn al-ʿAbbas, 57
Makkah, xviii, xxvi, 41, 43, 60, 85, 100
Maldives, xxiv
ʿAbd al-Malik, Abū Marwān, 76, 103
al-Maʾmūn, xxii, 33, 34, 44, 45, 47, 51, 52, 76, 88, 93, 96–98
al-manāẓir. *See* optics
Manichaean, 40, 48, 97
manṭiq. *See* logic
Marc of Toledo, xxv
Maʾrib, 42
al-Marwarūdhī, Khalid ibn ʿAbd al-Malik, 47
ibn Māsawayh, Yūḥanna, 33
Abū Maʿshar, Jaʿfar, xix, 16, 19, 51–53, 55
Māʾ al-Shaʿyr, 26
Māsirjawyh al-Ṭabib, 80
al-Masʿūdī, Abū al-Ḥasan, 4, 23, 26, 31, 85, 92
Matāriḥ Shuʿaʿāt al-Kawākib, 19
mathematics, xvii, xxii, xxiii, 18, 36, 59, 60–66, 69, 74, 75, 77, 80, 96, 99, 100
Mecca. *See* Makhah
medicine, xii, xxii, xxiii, xxiv, xxv, 16, 26, 30, 33, 34, 36, 37, 44, 47, 49, 55–57, 60–62, 65–68, 71, 73–78, 80, 81, 89, 90, 92, 96, 100
Medieval, xvi, 54, 86, 90, 91, 98, 105
Mediterranean Sea (Roman Sea), 20, 35, 58, 84, 91, 94
Mercury, 11, 12, 60, 86
Mesopotamia, xxiii, 20, 84, 86, 98, 101, 104
metaphysics, 48, 89
meteorology, 24
Meton, 27
Middle Ages, xv, xxii, xxiii, 81
Middle East, xviii, xxvi, 76, 90
al-Miṣri, ibn Yūnus, 55
Montpellier, xxiv
Morocco, xxiv
al-Muʾadhdhin, ʿibn Sulaymān, 60
Muʿawiyah ibn Abū Sufyān, 44, 60
Mūḍar, 4
al-mufradah, 81
Muḥammad (the Prophet), xxv, xxvi, 40, 43, 44, 48, 80, 82, 96, 100
ibn Muḥammad, Abū al-ʿArab, 75
al-Muḥyt al-Aʿẓam, 71

Muʿizz ad-Dawlah, 34
al-Mukhaṣṣaṣ, 71
al-Mumtaḥan, 51
al-Munṣūr ibn Abī ʿĀmir, 62, 63, 75
al-Munṣūr, Abū Jaʿfar, 33, 44–47, 96
ibn Abū Munṣūr, Yaḥyā, 47, 53, 55, 97
ibn al-Muqaffaʿ, ʿAbd Allah, 14, 45, 88
Murcia, 59, 77
Murūj al-Dhahab wa Maʿādin al-Jawhar, 85, 106
Mūsā, Banū, 51, 99
ibn Mūsā, Qāsim, 60
al-Muʿtamid, 52, 98
al-Mutawakkil, 33
al-Muʿtazilah, 61, 92

al-Najāt, 89
Najd, 4, 42, 68
ibn Najm, Abū al-Qāsim, 75
al-Nayryzī, al-Faḍl ibn Ḥātim, 28, 52
Naẓm al-ʿIqd, 46
Nebuchadnezzar, 18, 28, 90
Nicomachean Ethics, 24
Nicomack, 21, 23
Nihawand, 17
Nile, xxi, 35, 86, 94
Nimrod, 18, 89
Nisbat al-Akhlāṭ, 26
Noah, 4, 5, 15, 16, 89
Noble Reputation, The, 50
Nubians, 6, 8, 32, 86
numerals, 99
al-Nūn, Banū, 59, 68, 75, 76, 78, 83
ibn Nuṣayr, Mūsā, 83
nutrition, 33, 80

On Generation and Corruption, 24
On the Heavens, 24
optics, xxi, 23, 26, 27, 98
Orient, xi, 46, 83

Panchatantra, 88
Paris, 84, 105, 106
pedestrians, 22, 29
perigees, 13, 16
Peripatetic, 22, 29
Persia (al-Furs), xix, 3, 6, 11, 13, 15–18, 20, 25, 28, 41, 42, 44, 46, 47, 53, 55, 58, 64, 87–90, 94, 96–99, 105
Phaedo, 22
Pharaoh, 41
Philippus, 20, 25, 92
philosophy, xvii, xxi, xxii, xxiv, 6, 8, 19, 21–26, 29, 30, 32, 33, 36, 39, 42, 44, 45, 47–52, 56, 58, 62, 63, 66, 68, 69, 73, 76, 77, 79, 81, 82, 89, 99, 101, 103, 106
physics, 21, 23, 81, 89, 93, 99
planet, xxiii, xxv, 5, 12, 13, 16, 18, 37, 39, 40, 42, 43, 47, 51–54, 64, 66, 67, 72, 84, 86, 90
planetarium, xxiii
Plato, xxi, 21–23, 29, 30, 45, 48, 50, 92
Plato of Tivoli, xxv
poetry, xvii, 41, 55, 61, 63, 68, 71, 73, 81, 96, 100, 102
politics, 23, 24, 46, 50, 77
polytheism, 12, 25, 30
Porphyry, 25, 92
Posindrinus, 27
Proclus of Alexandria, 37, 95
prophecy, 22, 30, 40, 44, 48, 79, 87
Ptolemies, 20, 27, 91
Ptolemy, Claudius, xiii, 19, 27, 28, 37, 45, 47, 52, 53, 64, 84, 93, 100
Pyramids, 36, 94
Pyrenees, xxiv
Pyrrhon, 29
Pythagoras, xxi, 19, 21, 22, 25, 29, 30, 90, 92

Qaḍib al-Dhahab, 19
Qādisiyah, 17
Qaḥṭān, 38, 43
al-Qānūn (the Cannon), xxiii, xxiv, 27, 89
al-Qarshi al-Afṭas al-Marwānī, 66, 101
al-Qayrawani, Aḥmad ibn Ibrahim, 56
al-Qiblah, 60, 100

al-Qifṭī, xvii, 90, 92, 93, 96–98, 102, 103
ibn Qisṭār, Isḥaq, 81
Qur'an, 18, 40, 66, 76, 84, 89, 91, 96, 100, 105
ibn Qurrah, Abū al-Ḥasan, 33
ibn Qurrah, ibn Sinān ibn Thābit, 33, 74, 99
Qurtūba. *See* Córdoba
al-Qūṭ. *See* Visigoths

Raby'ah, 4, 40, 43
'Abd al-Raḥmān, Abū al-Muṭarraf, xi, 59, 68, 76
'Abd al-Raḥman, ibn al-Ḥakam ibn Hishām, xxii, xxiii, 55, 59
al-Rashid, Harūn, 33, 48, 55, 88, 91, 98
Ratio of Mixtures, 26
al-Razi, Muḥammad ibn Zakariyā, xxi, xxii, 30, 49, 56, 89, 97, 99
Red Sea, 42, 86
Republic, The, 23
rhetorical, 48
Richter-Bernburg, L., xviii, 84
ibn Abū al-Rijal, al-Ḥasan, xxi
Robert of Chester, xxv
Roger of Hereford, xxv
Roman Yarindj, The, 37
Rome (Roman), xix, xxi, xxii, 4, 6, 11, 20, 27, 28, 31–33, 35, 41, 45, 58, 59, 72, 79, 87, 90, 92–94
Rosenthal, Franz, xviii, 106
Rosetta (al-Rashid), 94
Ruba'iyat al-Khayyam, 89
al-Rūḥāni, 23
ibn Rushd Abūl-Walīd Muḥammad (Averroës), xxii, xxiv
Rūshum, 37, 95
Russia (Russian), xxiv, 4, 6, 24, 31, 32, 86

ibn al-Ṣabbāḥ, al-Ḥasan, 52, 98
Sabians, xviii, 5, 12, 17, 21, 29, 30, 31–33, 35, 40, 58, 74, 84, 86, 87
al-Saffāḥ, Abū al-'Abbās, 33
Abū al-Ṣaffār, Abū Ja'far, 67
ibn al-Ṣaffār, Abū al-Qāsim, 64–66
al-Sahlī, Ibrahim, 69
Saif al-Dawlah, 51
al-Ṣalafī, Abū Ṭāhir, xviii
ibn Sālim, Sa'īd, 72
al-Samā' wa al-'Ālam, 24
Samarkand, 3, 38
ibn al-Samḥ, Abū al-Qāsim, 64, 66, 101
Samī' al-Kiyān, 23
ibn al-Saminah, 60
San'ā', 4
Sarā'ir al-Ḥikmah, 18, 54
al-Sarkhasi, Aḥmad ibn Muḥammad, 49
al-Sarqasṭi, 67
Sarton, George, xviii, 85, 91, 98, 101, 107
Sassanids, 15, 16
Satan, 17
Saturn, 11, 12, 53, 60, 86
al-Ṣaydalāni, 'Ali ibn Khalaf, 69
ibn Sayyid, 'Abd al-Raḥmān, 69
science, xi, xii, xvii–xix, xxi, xxii, xxv, 7, 9, 11, 12, 15, 16, 18, 19, 20, 21, 22, 23, 25, 26, 28, 30–32, 35–38, 42–44, 46–56, 58, 60–73, 75–82, 84, 86–91, 93, 94, 96, 99, 100, 103, 106, 107
Seasons, The, 26
Seville, xxiii, xxv, 58, 59, 66, 67, 76, 101
sexagesimal, 91
al-Shahāb wa al-Haram, 24
ibn Shahr, Abū al-Ḥasan, 66, 101
Shajrat al-Ḥikmah, 63
ibn Shākir, Mūsā, xxii, 51, 98, 99
al-Sham, 4, 20, 21, 31, 39, 41–44, 47, 57, 68, 79, 84, 85, 89, 96, 97
al-Shifā', 89
ibn Shruyah, Abū 'Āmir, xvi
Sibawayh al-Baṣrī, 28
Sicily, xxiv, 64, 73, 93
al-Ṣiddyq, Abū Bakr, 43
al-Ṣiḥat wa al-Ṣuqm, 24
Simplicius, 27, 94
ibn Sīnā. *See* Avicenna

Sind, 4, 41
Sirius, 40
al-Siyāsah, 24
Slavonians, 7, 33
Socrates, xxi, 19, 21, 22, 30, 90, 92
Solomon, son of David, 21, 40
sophistical, 24
Spain, xi, xiii, xviii, xxii, xxiv, 83–85, 101, 103, 105
Stephen of Antioch, xxv
Ṣuār Darajāt al-Fulk, 16
Sudan, 6, 32
Sufi, 92, 102
al-Ṣulayḥi, ʿAli ibn Muḥammad, 67, 68, 101
ibn Sulaymān, Isḥaq, 80
Sumer, 90
al-Surqasṭī, Abū ʿUthman, 63
ibn Sydih, Abū al-Ḥasan, 71
syllogism, 24
al-Syrah al-Fāḍilat, 50
Syria(n), 42, 84, 85, 92, 93, 97
Syriac, 4, 94, 98, 99

al-Ṭabarī, Abū Jaʿfar, 70
al-Ṭabarī, ibn Rabban, xix, 56, 99, 102
al-Ṭabarī, ʿUmar ibn al-Farkhān, 33, 51, 55
Ṭabiʿat al-ʿAdad, 64
Taʿdil Zij al-Khuwarizmi, 53
al-Tafsyr, 16
Taghlib, xi
Tahāmah, 4, 43, 68
al-Tajibī, Abū Isḥaq, 68, 83, 102
al-Tajibī, al-ʾAmir, 68
al-Tajibī, ibn Ḥayy, 65, 67
al-taksyr, 74
Taʾlyf al-Luḥūn, 26
Tamim, 40
al-Tamimī, ibn Abū Rimtah, 44
ibn Tamlyḥ, Muḥammad, 73
Abū Tammam al-Ṭāʾy, 38, 95, 96, 102
ibn Ṭanj, Muḥammad, 33
al-Tanukhī, ibn Ismāʿīl, 52, 98
Taqdimat al-Maʿrifat, 26
al-Taʿrif bi Ṣaḥiḥ al-Tārikh, 57
al-Taʿrif bi-Ṭabaqāt al-ʾUmam, xv
Tārikh al-ʾUmam, xv
ibn Ṭāriq, Yaʿqūb, 55
al-Ṭaṣrif, xxiii
al-Ṭawāʾif, 15, 58, 62, 88
Tetrabiblios, 27, 52
ibn Thābit, Thābit ibn Sinān, 34, 74
Thales al-Malṭi, 25, 29
Thamūd, 38
al-Thaqafy, al-Ḥāryth ibn Kaldah, 44
al-Thaqfī, Abū ʿAbd al-Malik, 74
Themistios, 25
Theodosius, 27
theology, xii, xix, 12, 18, 21, 23, 24, 30, 36, 49, 50, 68, 69, 71, 77, 80, 96
Theon of Alexandria, 37, 51, 95
Thimār al-ʿAdad, 64
Thimār al-Ḥikmah, 13
Thimār ʿIlm al-ʿAdad, 64, 101
al-Ṭibb al-Ruḥani, 30
Timaeus of Physics, 23
Timaeus of the Spirit, 23
Timalaus, 27
Titus, 79
Toledo, xxv, 58, 59, 63, 68, 69, 74–78, 81, 98, 101, 103
Ṭulayṭilah. *See* Toledo
al-Ṭunayzī, Abū al-Qāsim, 63
al-Turjumān, Ḥūnayn ibn Isḥaq, 29
Turks, 4, 6, 7, 9, 11, 32, 45, 96
Tyre (Ṣūr), 26, 46, 59, 93

al-ʿUbūr, 40, 79
Udem, The Ethics of, 24
al-ʾUlūf, 19
ʿUlūm al-Burhān, 26
ʿUmar ibn al-Khaṭṭāb (al-Fārūq), xxvi, 17, 43, 80
Umayyad, xxii, 44, 58, 59, 62, 77, 101, 103
ibn Abū ʿUṣaybiʿah, xvii

Valens, 19, 37
Vāyu-Purāṇa, 87
Venus, 59, 60, 86
Vishnu-Purāṇa, 88

Vishtaspa, 17, 89
Visigoths, 58

Ahl Wabar, 39
Wālys, 19, 37
al-Waqshi, xii, 68, 83, 102
al-Wāsṭī, Abū al-ʾAṣbagh, 66
al-Waṣyfī, 36, 94

al-Yaḥsubi, ʿAbd Allah ibn Muḥammad, xviii
al-Yahudi, ʿIbri, 98
ibn Yaḥyā, Sulaymān (Jubayr), 81
ibn Yaḥyā, Yaḥyā, 60
Yathrib (Medina), xxvi, 43, 96
Yemen, 4, 40–44, 67, 68, 84, 101
ibn Yūnus, Abū Bushr Matta, 50, 71
ibn Yūnus, Aḥmad, 74
ibn Yūnus, ʿUmar, 74, 75, 102
Yūsuf, Abū al-ʿArab, 75

al-Ẓaffir, Ismāʿīl, xi, 83
al-Ẓāhir, 92, 102
ibn Zuhr, Abū Marwān, 76, 103
al-Zahrawī, Abū al-Ḥasan, 64, 65
al-Zahrawī, Abū al-Qāsim (Abulcasis or Albucasis), xxiii
al-Zarqāli, Abū Isḥaq ibn Yaḥyā, xvii, xxiii, 69, 84, 102
Abū Zayd, Ḥūnayn ibn Isḥaq, 33
ibn Zayd, Rabyʿ, 75
Zij al-Ḥaḍārāt, 88
Zij al-Shāh, 97
Zinj, 4, 6, 8, 12, 35, 42
ibn Ziyād, Ṭāriq, 83
Ziyādat Allah ibn al-Aghlab, 56
zodiac, xxv, 13, 39, 42, 46, 51
Zoroaster (Zaradusht), 16, 17, 88